Life and the
Universe
Exploring Eternity

By Walter Parks
And John Long

Copyright 2013
UnKnownTruths
Publishing Company

UnKnownTruths
Publishing Company
8815 Conroy Windermere Rd. Ste 190
Orlando, FL 32835
UnknownTruths.com
info@UnknownTruths.com

Contents

Foreword

"We go about our daily lives understanding almost nothing of the world. We give little thought to the machinery that generates the sunlight that makes life possible, to the gravity that glues us to an Earth…. or to the atoms of which we are made and on whose stability we fundamentally depend…. few of us spend much time wondering why nature is the way it is; where the cosmos came from, or whether it was always here.?"

Carl Sagan
From A Brief History of Time
By Stephen Hawking

When we do stop to observe and study our surroundings we are filled with awe and wonderment. We are compelled to try to understand more about our lives and the lives around us. We are driven to study and try to understand everything in the universe. With each new bit of knowledge we feel forced to learn even more.

Our wonderment over time has resulted in our animalistic brains producing complex minds that distinguish us from all other creatures on Earth. Our brains and our minds developed "side by side" during the development of the human race.

Our human brain and mind is the only thing on Earth that studies itself.

Maybe you have pondered the questions of life and the universe as the authors of this book have.

When you think about it, life is much more complex, yet much more understandable than when we were kids.

We all start out believing what we are taught, but our beliefs evolve as we go through life and are exposed to new events and new people with new ideas. Many of us then become curious enough to actively seek new knowledge.

What we believe as we get older is a product of this life long process. But what we believe always remains biased by the teaching of our youth.

In this book we have tried to remove all youthful biases and document what we really believe today. But we, and you, can sense a bit of lingering bias.

We tackle some of the biggest questions of mankind with the confidence that we can arrive at the truth and convey that truth in comprehensible narrative.

We hope you enjoy our efforts and even if you cannot agree with all of our conclusions we hope this book will stimulate you to continue to seek the truth about everything that is important to you.

We should acknowledge that our research relied heavily on the Internet and the many conflicting sets of information one can find there. We had to do a lot of sorting, but we are grateful to all of those who laid out their thoughts and beliefs on that greatest of all information sources. We thank them all.

Introduction
Life and the Universe
Exploring Eternity

A mind that is stretched by new experience can never go back to its old dimensions.

Oliver Wendell Holmes

Just what is life? What do we really know about God? What do we really know about the universe? Is there intelligent life out there? Are we likely to encounter aliens in our lifetime?

Is there more than one universe? Will parallel universes be proven beyond a shadow of a doubt?

These are just some of the questions that two friend have been asking since we were in grammar school together over 70 years ago. From our earliest days together at school we realized that we had a very special relationship with a great hunger for knowledge, and a thirst for answers. We enjoyed exploring each other's minds.

We have continued to speculate with each other about life, the universe, and the great unknowns that cause mankind to be such a special species here on earth.

We have been living in different states, Mississippi and Florida, since high school graduation. One of us became an aerospace engineer and later a Vice President at Lockheed Martin Aerospace Company. The other became a trial attorney and went into private law practice. But we still meet, write, email and discuss our evolving thoughts and ideas.

We recently decided to meet in Homosassa Springs Florida for a few days to chat about our ideas and see just what we really think about the answers to all the great questions of life and the universe.

We grew up with Christian backgrounds, one a Baptist and one a Catholic.

We were especially interested in discussing how our evolving thoughts and ideas have changed over the last 70 years.

And although we did not discuss it outright, it was obvious that we may have been motivated to have this meeting because of increased interests in seeking answers to the questions: do we have souls and if so, is there an afterlife.

We met in the wilderness of the springs to discuss and try to find the best answers available based on what mankind has learned to date and based on the life long questioning of two very curious old men.

We felt that our findings and understandings would be of general interest to a wide audience, especially the 79 million baby boomers in the United States. We also thought that young people who are just beginning to form their opinions of life and the universe would find our materials interesting and useful. So we wrote down our meeting results and published them in this book.

We began by staring at the wonders of the natural world there at Homosassa Springs in Florida. This world has been mostly out of sight for those of us who live in relatively populated country sides and cities.

The Wilderness Where We Met

One of us has continued to live in the country side of his birth in Saltillo, Mississippi and the other in the city of Orlando, Florida.

We stared into the night sky and marveled at the immensity of the universe. It had been a very long time since we were away from the lights of civilization and could see the heavens so clearly.

We saw an uncountable number of stars. It has been estimated that there are more than ten million billion billion stars in the universe. And there are most likely many more planets since it is believed that each star will likely average several planets.

We are awed at the enormity of our universe.

With so many planets it is easy to speculate that there are large numbers of intelligent beings in our universe.

We have walked in the forests and by streams and are awed at the great variety and complexity of life here on earth.

When we look into our deep oceans we see extreme life forms living even in the very hot volcanic vent flows from the ocean floor.

Life at Deep Ocean Hot Volcanic Vent Flows

Some of these life forms look like monsters from our imaginations.

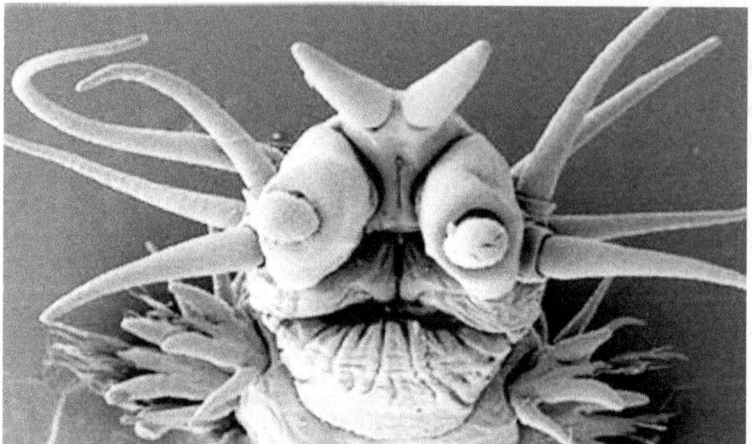

A Worm Living in Extremely Hot Deep Ocean Vent

We have found life in the extreme arctic conditions at the poles.

Undersea Life in Arctic Extreme

Life here on earth occupies all environments no matter how severe.

This suggests that we should therefore expect life on essentially all of the trillions of planets in our universe that have life acceptable temperatures.

We believe that we will soon learn that millions if not billions of these alien life forms are intelligent; and thousands if not millions are more advanced than are we. It's just a matter of statistics and the older age of so many of the planets.

We have speculated about life and the universe and other unknowns all of our lives. We are 77 years old as we write this.

Fundamental to all answers is the question of God.

We decided to explore Life and the Universe in the steps listed in the Table of Contents.

We wanted to explore and conclude what we really know in this age of vast informational knowledge being spread by the Internet. And, based on this knowledge, what can we predict for the future?

We hope that our efforts will, to paraphrase **Oliver Wendell Holmes**, *stretch your mind with a new experience and perhaps change your life.*

Chapter 1
What is Life

"Life is a moderately good play with a badly written third act."
Truman Capote

From the earliest days of becoming thinking humans we have had many questions about life. We have had fewer acceptable answers. This is beginning to change as we are learning more and more about life and our universe.

A series of modern day scientists and physics from Einstein to Hubble to Heisenberg to Kaku to Hawking have postulated theories in an attempt to answer our most complex questions. Their once unconceivable theories are rapidly becoming acceptable and practical as billions of dollars are now being spent to test and prove the validity of their theories.

Let's explore what we have learned about Life.

Life as We Know It

As we walk about Earth and look around we learn that there are about 10 million species of animals and several hundred thousand species of plants. But all lives that we see are latecomers on planet Earth.

And even the fossils of old that we find, when added to the life forms of today make up only a small fraction of species that have lived on planet Earth.

The real answer to "What is Life" lies in the greater diversity of microorganisms. Bacteria, protozoans and algae make up by far the most life on Earth.

When you consider the ecological circuitry of Earth, i.e. the ways in which materials like carbon, sulfur, phosphorous and nitrogen get cycled in ways that makes them available for all biology, the organisms that do most of the real work are bacteria.

Bacteria are essential for our ecology; we are optional.

The big difference between us and a bacterium is that our body consists of trillions of cells that function in a coordinated manner. Bacteria are single cells, although they're not really independent. Bacteria actually live in associations of great numbers. They are not lone operators. They work in these very highly coordinated communities of organisms that help each other to grow and prosper.

We will discuss this view of life in later sections of this book. Let's first explore the "What is Life" question from the meaning of humans.

Life can be described in terms of religion, philosophy, and biology. Each of the three describes the conception and nature of life in different terms.

The Religious Views

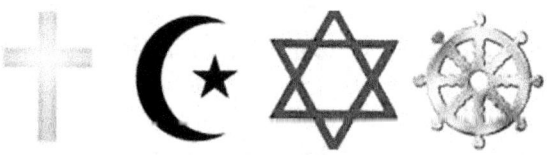

Many people live their lives without ever reflecting on life itself or its meaning for them. A few may feel "a calling" early on in their lives that help to focus their lives and their view of life.

But most of us never develop such "a calling" and live our lives with no overall purpose. We may not develop a clear idea of our own nature or identity. We frequently have no idea of who or what we really are.

However, if you question us and "force" us to think a bit and force an answer, it will most likely be religiously based.

Some early men felt the need to try to find a purpose in life and understand what life may be. They speculated and from their speculations many religions were developed.

So when we are forced to say what life really is we usually turn to these old religious teachings.

Life is a gift from God.

Our human lives are special because we are made in God's image.

We were made by God to share in God's own life.

Muslims generally believe that Allah made us so we could worship Him.

Christians and Jews generally believe that God made us to be caretakers for the earth and to share a relationship with Him.

Others are still searching to try to find man's responsibilities to God and the spiritual meaning of life.

Most religions teach that only the true religion can give purpose to human existence. It is only our relationship to the Creator, and the purpose which that Creator has fixed for His creatures, that human existence has any meaning.

Atheists do not believe in a Creator. They believe that humans are simply chance products of the thermodynamic system of physics; we are just happenstances. We are just "temporary material existences of conscious animals moving through life". We are driven by animalistic instincts and seek as much pleasure and as little pain and suffering as possible.

Religious leaders down through the ages have attempted to "educate" everyone, even including the atheists, to the fundamental and spiritual role of religion. They have tried to enable people to achieve a true understanding of their own nature and of God's will and purpose for them.

They tried to teach us that our daily lives should be focused toward our spiritual development. We must focus on our soul because it is the only part of us which endures. Whatever promotes our spiritual development is good, and whatever hinders it is bad.

Most religions teach us that God has a plan for the universe and that each of our lives is meaningful only to the degree that we help God realize this plan; fulfilling God's purpose is the sole meaning of life and it is the only way we can receive an afterlife with God.

But some believe that even if there is no spiritual realm, meaning in life is possible. They believe that a significant existence can be had in a world known only by science.

They believe that something is meaningful for a person if he/she believes it to be meaningful. Life can be meaningful just because of its intrinsic nature.

Life is more meaningful when we get what we want; when we achieve a highly desired goal; when we really care or love someone or something. **Life is meaningful when you believe that your life is meaningful.**

Some atheists however, have negative feelings about the meaningfulness of life. Because they do not believe in the existence of God or a soul or a divinely structured universe, they cannot see any way that life could be meaningful. They just accept the "fact" that the universe and all of us are just happenstances.

They question how we "trivial" beings could have an impact on the world. What could we do in our short lifespan that could affect the universe that is billions of years old?

It is clear that our religions at best offer only reasoned answers that have no basis in fact. Most of the answers were derived centuries ago by religious leaders desperately seeking the truth. So the answers based on religion generated in those ancient times are rather contrived.

Religions have not given us a clear answer to "What is Life"; we have to look elsewhere for the answers.

The Philosophers View

All views of "what is life" lead to the questions, "is there a God" and the related question of "do we have free will".

The most definitive information regarding the views of philosophers is from an older survey of the members of America's elite National Academy of Sciences back in 1998. This survey asked the members to give their views of God and of freedom.

Only 7% believed in God.

About 12% of philosophers think that people's lives are predestined. What does this mean regarding free will?

Nearly three-quarters of the philosophers accept or lean towards atheism.

Some 82% of the respondents accept or are inclined towards "non-skeptical realism" about the external world, which means they believe both that physical objects exist independently of the minds that perceive them, and that we can be said to know of their existence.

Some 4.8%, though, are inclined to deny that we have certain knowledge of the existence of physical objects, and 4.2% accept or lean towards "idealism", which is the theory that matter somehow depends on mind.

As to "abstract" objects, 39% believe in "Platonism", that is, they believe that abstract objects have a real existence independently of our minds.

Today's philosophers believe that judgments of artistic value are not merely matters of individual taste: 41% said aesthetic values are objective, 34% say subjective, and a quarter gave some other answer.

Some 56% incline towards "moral realism", which has no precise definition but implies that ethical questions have objectively right and wrong answers.

Nearly two-thirds endorsed moral "cognitive existence", which suggests that they believe there are moral facts or truths.

In other questions 15% or more of the philosophers said they were too unfamiliar with the issue to give an opinion.

Philosophy is now a highly specialized discipline and many believe that philosophy has given up on dealing with the big questions of life and is now mired in technical minutiae.

But even Plato was attacked in his own time for treating philosophy as if it were all mathematics. And 1,800 years ago the great Doctor Galen moaned about "the over-refined linguistic quibbling of some philosophers".

Socrates believed in the immortality of the soul and that the gods had singled him out as a divine emissary.

Socrates Teaching

Socrates questioned the doctrine that virtue can be taught. He liked to observe that successful fathers did not produce sons of their own quality. Socrates argued that moral excellence was more a matter of divine bequest than parental nurture. This belief may have contributed to his lack of anxiety about the future of his own sons.

Socrates believed wrongdoing was a consequence of ignorance and those who did wrong knew no better.

The one thing Socrates consistently claimed to have knowledge of was "the art of love", which he connected with the concept of "the love of wisdom", i.e., philosophy. He never actually claimed to be wise, only to understand the path a lover of wisdom must take in pursuing it.

Even when the jury had sentenced him to death, Socrates calmly delivered his final public words which were a speculation about what the future holds. Disclaiming any certainty about the fate of a human being after death, he nevertheless expresses a continued confidence in the power of reason.

Plato followed Socrates as his principle and best known student.

Plato's hypothesis that our soul was once in a better place and now lives in a fallen world made it easy to combine platonic philosophy and Christianity, which accounts for the popularity of Platonism in Late Antiquity.

Plato Making a Point

Plato's most famous student was Aristotle. But Aristotle replaced his master's speculations with a more down-to-earth philosophy which is described in a series that have become classics.

Most philosophers believe that Aristotle is the most influential philosopher of all ages and was the founder of modern science.

Aristotle

Eventually philosophers from all the various schools borrowed concepts and ideas from other branches of philosophy. The various schools of philosophy began to merge to a new synthesis (called Neo-Platonism) which was created by Plotinus (205-270). Plotinus believed that our world was a mere shadow of an even higher world which was a shadow of the One God.

That is, he believed that the world has four levels of reality: God was the highest level, and then there were the levels of the intellect, the soul, and matter.

According to Plotinus, the wise man would try, by means of virtue, to free his soul from matter and unite it with God.

His philosophy was adopted by the fathers of the Catholic Church, Ambrose and Augustine, and was to remain the philosophical school par excellence until Aristotle was rediscovered in the twelfth century.

So what answers to the question "what is life" do we get from the Philosophers? We get nothing really; at least nothing understandable for our purposes.

Most of the great ancient ones did believe in God but the vast majority of those today do not believe in God.

The ancients and the new all just philosophize. They provide no real answers to the great questions of mankind.

We have to look elsewhere for any concrete answers.

The Biological View

It is perhaps best to rely on the biological view of life where we have scientifically defined information.

Life distinguishes objects that have self-sustaining biological processes from "inanimate" objects. Life is the condition which distinguishes active organisms from inorganic matter.

Living organisms here on earth have carbon and water-based cellular forms with complex organization that contain heritable genetic information.

Living organisms undergo metabolism, possess a capacity to grow, to respond to stimuli, to reproduce and, through natural selection, adapt to their environment in successive generations.

The more complex living organisms can communicate through various means.

Earth's life forms vary in complexity from one cell creatures such as an amoeba to the most complex which is man.

The cell is the basic unit of a living organism.

The cell is the simplest unit of matter that is alive. From the unicellular bacteria to multi-cellular animals, the cell is one of the basic organizational principles of biology. There are basically two types of cells: prokaryotes and eukaryotes. Prokaryote cells have no nucleus and are the basic life form for some bacteria and algae.

Prokaryote Cells

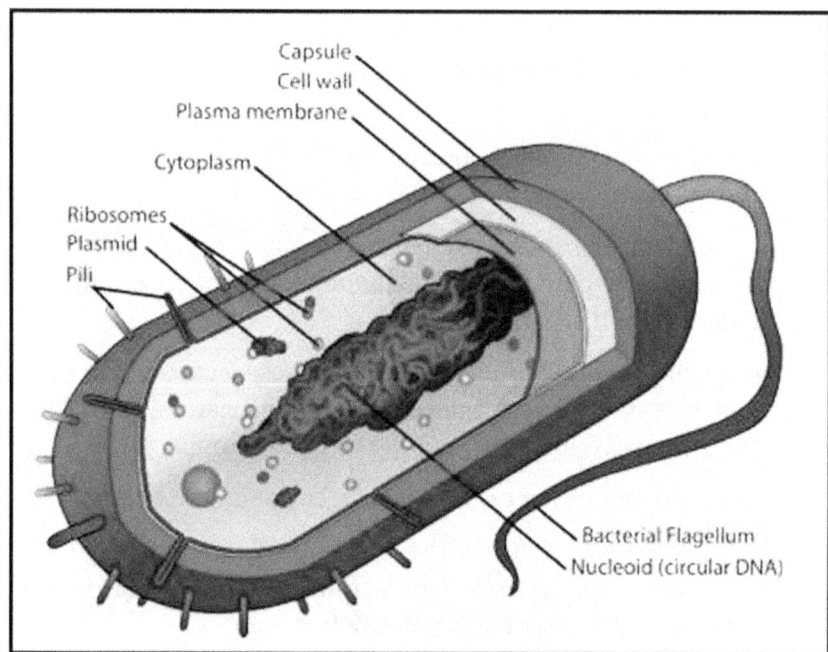

Prokaryotic Cell Structure

Prokaryotes are single-celled organisms that are the earliest and most primitive forms of life that are still on earth. Prokaryotes include bacteria and related micro organisms. Prokaryotes are able to live and thrive in various types of environments including extreme habitats such as hydrothermal vents, hot springs, swamps, wetlands, and the guts of animals.

Prokaryotic cells are not as complex as eukaryotic cells. They have no true nucleus as the DNA is not contained within a membrane or separated from the rest of the cell, but are coiled up in a region of the cytoplasm called the nucleoid. Using bacteria as our sample prokaryote, the following structures can be found in bacterial cells:

Capsule - Found in some bacterial cells, this additional outer covering protects the cell when it is engulfed by other organisms, assists in retaining moisture, and helps the cell adhere to surfaces and nutrients.

Cell Wall - Outer covering of most cells that protects the bacterial cell and gives it shape.

Cytoplasm - A gel-like substance composed mainly of water that also contains enzymes, salts, cell components, and various organic molecules.

Cell Membrane or Plasma Membrane - Surrounds the cell's cytoplasm and regulates the flow of substances in and out of the cell.

Pili - Hair-like structures on the surface of the cell that attach to other bacterial cells. Shorter pili called fimbriae help bacteria attach to surfaces.

Flagella - Long, whip-like protrusion that aids in cellular locomotion.

Ribosomes - Cell structures responsible for protein production.

Plasmids - Gene carrying, circular DNA structures that are not involved in reproduction.

Nucleiod Region - Area of the cytoplasm that contains the single bacterial DNA molecule.

Most prokaryotes reproduce through a process called binary fission. During binary fission, the single DNA molecule replicates and the original cell is divided into two identical daughter cells.

Eukaryote Cells

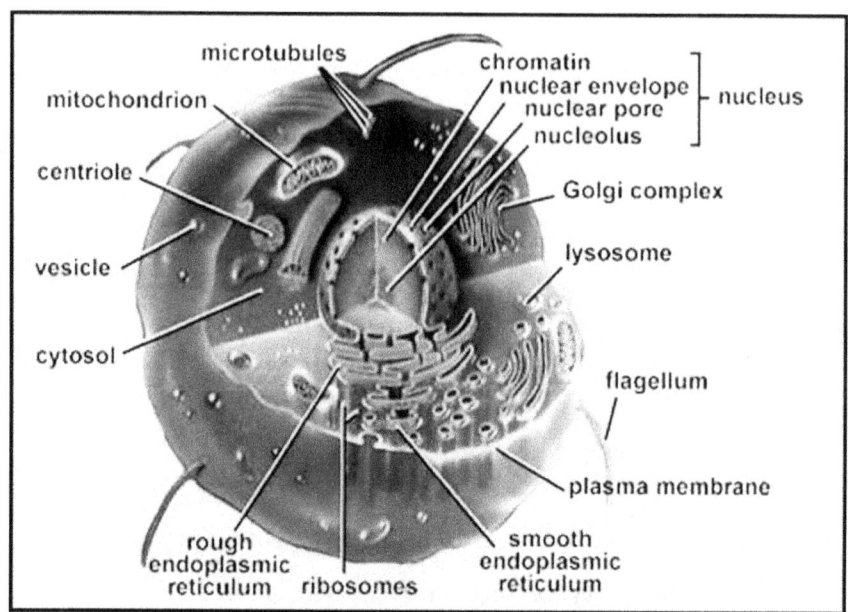

A model of a eukaryotic cell
(Picture taken from On-Line Biology Book)

A eukaryotic cell has a nucleus, which is separated from the rest of the cell by a membrane. The nucleus contains chromosomes, which are the carrier of the genetic material.

There are internal membrane enclosed compartments within eukaryotic cells called organelles, e.g., centrioles, lysosomes, golgi complexes, mitochondria among others as shown above. Each is specialized for particular biological processes. The mitochondria are found in all eukaryotes and are specialized for energy production.

The area of the cell outside the nucleus and the organelles is called the cytoplasm.

Membranes are complex structures and they are an effective barrier to the environment, and regulate the flow of food, energy and information in and out of the cell.

There is a theory that mitochondria are prokaryotes living within the eukaryotic cells.

The eukaryote cells have a nucleus and are the basic life form of the higher order of creatures.

All cells have the same basic structure. They have an outer covering called a plasma membrane. The plasma membrane holds the cell together and permits the passage of substances into and out of the cell.

The interior of cells is called the cytoplasm. Within the cytoplasm of eukaryotes are embedded the cellular organelles which perform the functions of the cells.

In organisms with more than one cell, a collection of cells that work together to perform similar functions is called a tissue. Tissues that perform coordinated functions form organs. Organs that work together to perform general processes form body systems.

There are many types of cells. Multi-cellular organisms contain a vast array of highly specialized cells. Plants for example contain root cells, leaf cells, and stem cells. Examples of human's cells include skin cells, nerve cells, and sex cells. Each kind of cell is structured to perform a highly specialized function.

All life on planet earth is designed and operated by its DNA which is a very large molecule in the nucleus of each cell. The DNA caries codes that perform 3 functions:

- Builds the proteins that structures our bodies;
- Operate switches that turn the protein making genes off and on;
- Operate the "command center" that activates the switches.

We can better understand this biological nature of life by understanding the DNA structure of our chromosomes and telomeres.

DNA Structure of Chromosomes and Telomeres

DNA is a double stranded helix made up of base pairs. It is frequently depicted as follows:

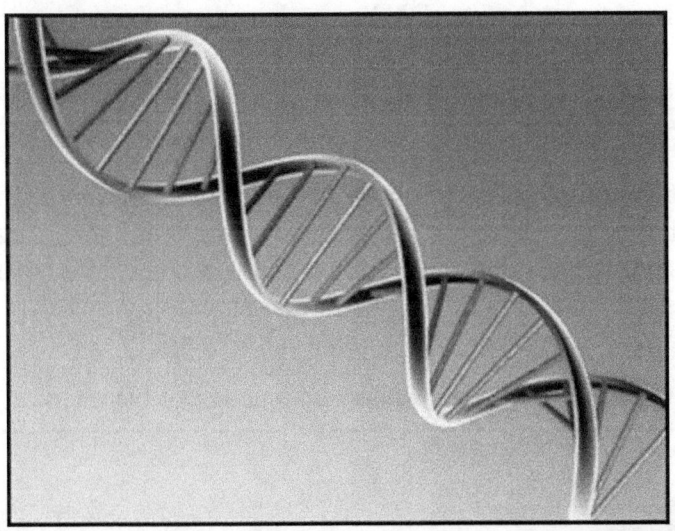

We need to take a closer look to better understand the nature of the base pairs.

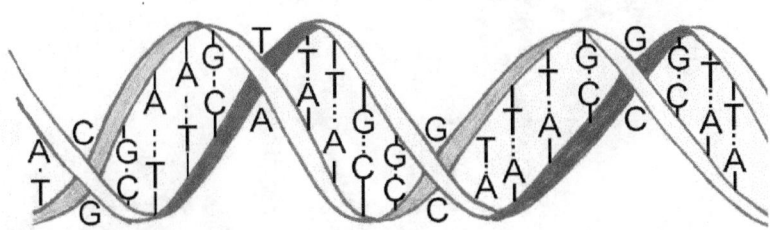

The base pairs, i.e. the "steps in the DNA ladder" are made of 4 substances: adenine on one strand (represented with an "A") always pairs with thymine (represented with a "T") on the other strand. These are called A-T pairs regardless of which strand has the "A" and which the "T."

Similarly, cytosine on one strand (represented with a "C") always pairs with guanine (represented with a "G") on the other strand, creating G-C pairs.

Scientists often represent DNA strands with a string of letters like this:

ATATTTGAAAGCTGTGTCTGTAAACTGATGGCTAACAAA ACTAG.

This string of letters represents only one strand, or one half of the DNA molecule. There is no need to write down the other strand because as we just described above, a "G" in one strand means there is automatically a "C" in the other strand, just as a "C" in one strand implies that the other contains a "G".

A group of these sequences make up a gene as illustrated.

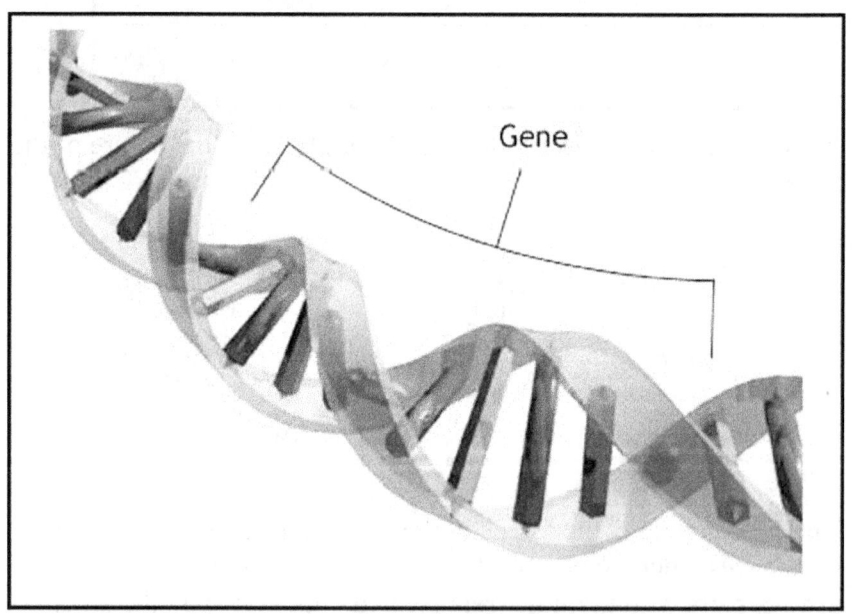

Gene

The average gene is about 27,000 base pairs long. The longest discovered is a muscle production gene that is 3.2 billion base pairs long.

At the end of each chromosome are telomeres which are non-coding repeating genes with the base pair sequence of TTAGGG. The telomere can reach a length of 15,000 base pairs.

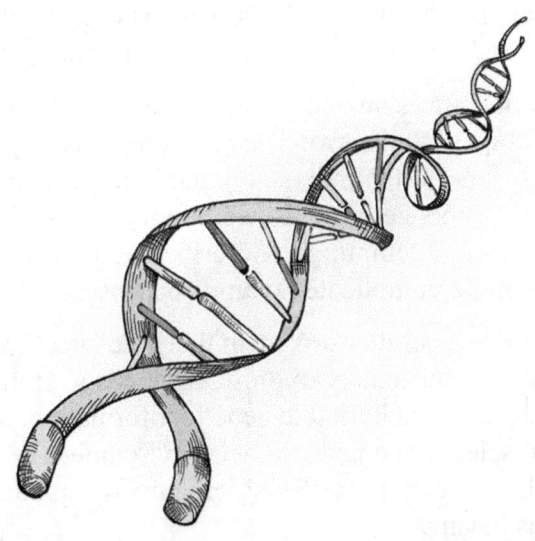

Telomeres Cap Chromosomes for Protection

Telomeres function by preventing chromosomes from losing base pair sequences at their ends. They also stop chromosomes from fusing to each other. However, each time a cell divides to replace itself, some of the telomere, usually about 25-200 base pairs per division, is lost.

The telomeres have been likened to the grommets at the ends of shoelaces that prevent them from fraying.

When the telomere becomes too short, the chromosome reaches a "critical length" and can no longer replicate. This means that its cell becomes too "old" and dies without being able to replace itself.

In a simplified description of the DNA operation, the sequence of base pairs that make up a gene split in the center exposing the ATCG bases. These bases attract materials from within the cell to replace the other side of the bases, e.g. TAGC in a structure called

RNA. This becomes a protein that the body needs for its structure and growth.

The protein is made from 20 amino acids that the DNA fines within the cell. The protein may be composed of a small number of amino acids or very large numbers of the 20 basic amino acids.

The genes that operate as switches and the genes that send "command messages" to the switches have only recently been found and are still not fully understood. Only recently these genes, since they did not produce proteins, were thought to be "junk DNA". We will skip attempting to describe them further less we make this book more complicated than its purpose warrants.

So biological life is a cellular organism that feeds itself, undergoes metabolism, possess a capacity to grow, responds to stimuli, reproduces and passes on heritable genetic information, and through natural selection, adapts to their environment in successive generations. The more complex living organisms can communicate through various means.

We now know that all life on earth evolved from very primitive DNA, as we will discuss in later chapters.

Life is the DNA that mutates and by so doing creates all the various species. The life of a specific species is the species DNA.

Each of us humans are the continuing and evolving life from the DNA "seed" planted or created on earth about 4 billion years ago.

So we see that DNA is the basic answer to our question: "What is life?"

Life is DNA. Yes, that's the basic answer, but not the final answer as we learn when we explore the next question: How did life begin?

Chapter 2
How did Life Begin

"I don't know the question, but sex is definitely the answer."
Woody Allen

OK, so we know that all life was created or evolved from a primitive DNA "seed". We can trace all life here on earth back almost 4 billion years ago to that first seed. But from where did that first "seed" come?

And how did that first seed lead to the great variety of creations now living on earth?

How did that first seed lead to the creation/evolution of man?

It's complicated. Let's begin by reviewing how non-life first became life.

Crystalline Growth

Let's start by considering crystals. **They grow via the laws of physics and chemistry.**

Snowflakes and ice crystals form as the temperature drops below that of liquid water; they **"crystallize"**.

They will continue to grow as long as liquid water is available and the temperature remains below the freezing point. They can also grow from sublimation from water vapor in the atmosphere.

In a similar manner crystals of other materials form and grow.

There are many materials that form growing crystals.

Crystals can grow via various processes including evaporation, a change in the pH (acidity) of the solution from which they grow, temperature changes, or other changes in the solution from which they grow.

When crystals form slowly, each ion or molecule finds its correct place in the crystal lattice and we get an almost perfect crystal. When the precipitate is very rapid, we can get crystals that are impure.

OK, so we see how **non-life may grow**. Now let's see how first life may have developed from such crystalline-like processes.

Creation of Amino Acids

In 1953, Miller and Urey speculated about what conditions may have existed on primordial Earth.

Stanley L. Miller and Harold C. Urey

They combined ammonia, hydrogen, methane, and water vapor and inserted electrical sparks to simulate energy provided by lightning.

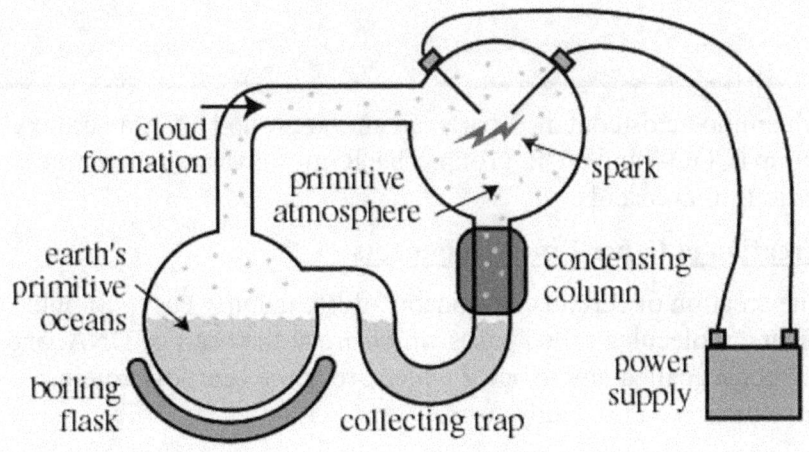

This experiment resulted in the formulation of new molecules that they identified as the molecules of eleven of the standard twenty amino acids that make up proteins of today's life. [It should be noted that a 21st amino acid has recently been found and a 22nd may soon be found.]

The typical amino acid structure is as follows.

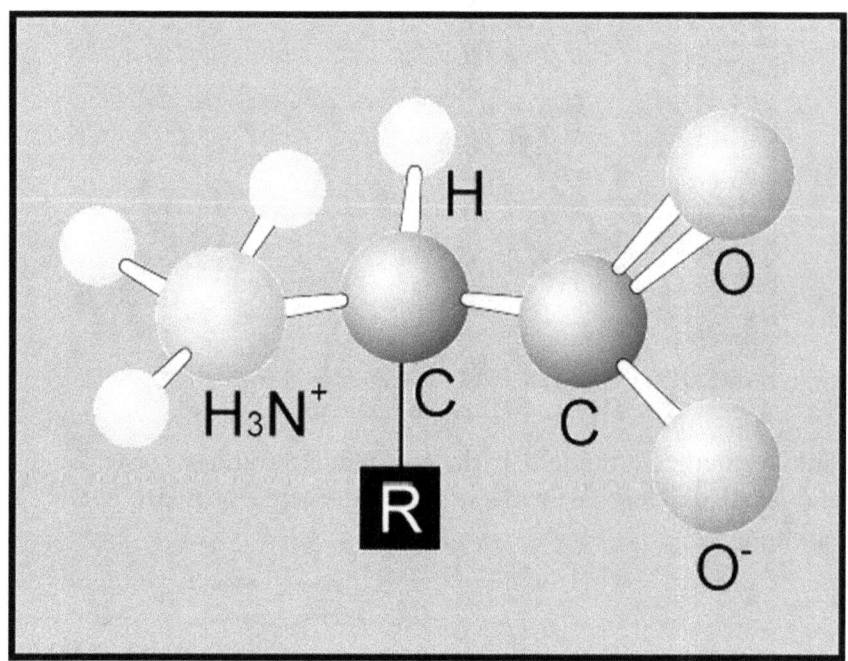

All amino acids contain 3 parts, an amino group H_3N+, a carboxyl group (COO-) and an 'R' group. The 'R' group varies among the various amino acids.

Creation of Other Life Components

The creation of certain components of life is fairly easy. Simple sugars, molecules called bases which are at the heart of DNA, and molecules called amino acids which are at the heart of proteins have been seen to readily form.

It's also fairly easy to make some of the fatty substances that make the coverings of cells.

Making all of these building blocks individually seems to be pretty plausible.

The harder part is how to get them to work together. How did it go from some warm pond on the primordial Earth that has amino acids, sugars, fatty acids, et al just floating around in the environment, to something in which nucleic acids are actually directing proteins to make the membranes of the cell?

We will try to describe how life began, giving our best answers, but we must recognize that the initial emergence of life was most likely an event that occurred 3.5 billion to 4 billion years ago. Even the rocks of that era have mostly vanished so we will have to rely on some speculation in fill in the gaps between actual evidence.

Life most likely started in the simplest possible way, as a cycle, i.e. a natural chemical reaction that repeated itself, spinning off byproducts. Some of these byproducts survived to develop and maintain the cycle.

But where and how did this cycle start? Dr. Wächtershaüser favors some mineral surface like iron pyrites, also known as fool's gold. The iron pyrites are natural catalysts that could have assembled chemicals like carbon monoxide into biological building blocks.

This beginning process is not unlike the crystal growth described above for non-life.

At some stage, the little cycle acquired a cover of protective chemicals to separate its own reactions from the general environment. When the cover eventually enveloped the cycle and broke free of the mineral surface, the first cell was born.

Recently an article in Science Magazine described how researchers from the Massachusetts General Hospital reported that montmorillonite clay, formed from weathered volcanic ash and familiar in many households as cat litter, has a property of possible relevance to the origin of life. It makes droplets of fat molecules rearrange themselves into small bubbles, similar to the membranes that make up the walls of living cells.

Often clay particles are incorporated into the bubbles, the research team found, and could contain attached pre-RNA and RNA molecules. They concluded that such mineral particles may have greatly facilitated the emergence of the first cells.

In a second experiment, the researchers found that they could make their proto-cells divide by forcing them through fine holes in a filter. A natural counterpart to this process, they suggest, would be water currents forcing bubbles through rock pores. (We will later discuss the importance of how cells "know" when to divide.)

The First Life

All of this tends to support the Miller and Urey conclusion that life's first organisms on Earth likely arose in an environment similar to the one they constructed in their lab. This "environment", rich in organic compounds, is now widely described as the **primordial soup**.

This hypothesis is further extended to the claim that, within this soup, single-celled organisms evolved, and as the number of organisms increased, the organic compounds were depleted.

Necessarily, in this competitive environment, those organisms that were able to biosynthesize their own nutrients from elements had a great advantage over those that could not.

[Today, the vast majority of organic compounds derive from biological organisms that break down and replenish the resources for sustaining other organisms. And, rather than emerging from an electrified primordial soup, amino acids now primarily emerge from biosynthetic enzymatic creations; and the remains of dead organisms.]

The Miller and Urey conclusions may well be how life got started at the cellular level with simple self-replicating molecules derived from the created amino acids.

The RNA World

The next steps of how these "created" amino acids came together to form combinations and then RNA which developed into DNA requires a very detailed technical discussion. We have omitted the technical discussion of these steps because it would take us down a tedious and detailed discussion of organic chemistry that would likely detract from understanding the **fundamentals** of how life began on Earth.

But we do need to note that biologists have long supposed that RNA was the key component in early cell development. It could act as an enzyme, a catalyst of chemical activities and could store genetic information. Most of the early RNA functions, including information storage were later passed to DNA which is a more stable molecule.

Some believe that this step by step creation from crystals to DNA is sufficient to describe how life began in that after a sufficiently long period of Darwinian evolution the humble non-life replicator cell eventually transformed into an entity complex enough that it became indisputably living.

These early biological systems can be distinguished from chemical systems because they contain components that have many potential alternative compositions; and that through trial and error evolution can adapt themselves and create a molecular memory or genotype which becomes shaped by experience, by Darwinian selection and are maintained by self-reproduction.

Suffice it to say that Miller and Urey pretty much proved that amino acids, the building blocks of life, can almost spontaneously form in the environment believed to have existed in the early days of earth.

We, the authors, accept this theory for single-cell simple life.

We also can readily accept the facts of cellular osmosis whereby water and nutrients can be absorbed from a "soup" outside the cell and that waste products can osmoses out of the cell. We therefore can see how cells could receive sufficient nutrients to grow.

A Void in Evolution's Answers

However, it is harder to see how the cells **"know"** when they have become large enough that they need to split and form a daughter cell. From where may they "get" that instruction?

Evolution theory does not seem to provide the answer.

The cells need information to tell them how to proceed to life, and in life.

So this raises a more basic question about creation/evolution; how did the first **informational molecules** come into existence, i.e. the molecules needed to **direct** the generation of efficient self-replication machinery.

The purely mechanical forcing of cells to divide as previously described may have initiated the creation of cell division instructors such as the special proteins called **cyclins** as described below.

Complex Life

However more complex life did not really begin until there was production of macromolecules that served as primitive stores of genetic information, i.e. until genomes that contained the informational molecules were created.

Life is more than just complex chemistry; it requires unique **informational management properties** that are crucial factors that escalate non-life to life.

The manner in which information flows through and between cells and sub-cellular structures is quite unlike anything else observed in chemical-only nature.

Functionality

This biological information flow gives each cell and the organism of which it is a part the **functionality** needed for life.

The information content of DNA is only a small part of the story. DNA is not a blueprint for an organism; DNA is a (mostly) passive repository for transcription of stored data into RNA, some (but by no means all) of which goes on to be translated into proteins.

The biologically relevant information stored in DNA therefore has very little to do with its specific chemical nature. It is the functionality of the DNA expressed RNA and proteins that is biologically important.

Functionality, however, is not a local property of a molecule.

It is therefore not possible to determine what will be functional in a cell based on local structure and sequence information alone. It is thus clear that functionality, the most important features of biological information is decisively non-local. Functionality is subject to informational control and feedback; it changes with time in a manner that is both a function of the current state and the history of the organism.

Biological systems information is a property of the system; it is distinctive because it is the information that determines the current state, its dynamics of change and the future state of the system.

Information control is the key defining characteristic of life.

But how did information molecules first come into being and gain control for living organisms?

Early systems without information control existed in the past but were not likely to evolve over geological timescales without acquiring informational protocols. Therefore, life forms with informational control may be the only systems that evolved in the long-run and are thus the only remaining product of the processes that led to complex life. [Some very simple "life" that did not gain the informational molecules did survive and some still "live" today. But that is not the kind of life we are interested in for this book.]

Thus the onset of Darwinian evolution in a chemical system was probably not the critical step in the emergence of life. Instead, **the emergence of life was probably marked by a transition to information processing capabilities.**

Of major interest is to determine how information control emerged in the pre-RNA and then a RNA world setting from just chemical kinetics to life forms with primitive information control mechanisms.

Let's look back at the simple one cell animal and consider how it gets its food.

It must be in a nutrient laden environment and the chemical (called partial pressures) pressures of the water and nutrients outside its body, i.e. outside its cell must be greater than the pressures inside. The water and nutrients then osmoses trough the cell wall into the cell so that it can grow.

The cell does not need informational molecules to allow this to happen.

But how does the cell know when to stop growing and to split into daughter cells?

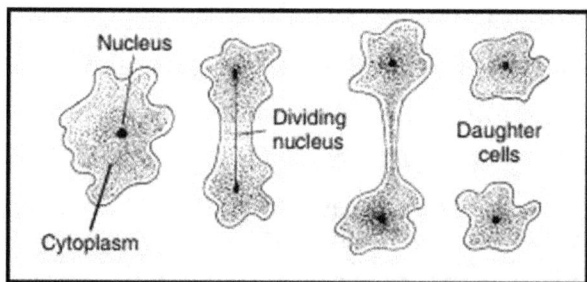

All cells follow a routine. Cells know (because of chemical pressure differentials between inside and outside) when they have to take in food, when to grow and in the same way they know when it is time for them to divide. This routine is called the cell cycle.

The life of a cell begins when a parent cell divides by mitosis. This process is very well controlled by the cell. When they are dividing, cells produce special proteins called **cyclins**. These proteins are produced only during a particular time of the cell cycle.

If the cell grows too large there will not be enough interface between the inside of the cell and its outside (called wetted area) to provide enough cell wall surface for sufficient nutrients to be absorbed and the cell will starve and die.

In the Darwinian evolution process over a long period of time some cells developed the ability to sense this situation and an informational molecule was created in their DNA that tells it to divide into daughter cells. This sensing led to the creation of the molecule that produces the messenger protein cyclins as mentioned above.

Such a process of creation for various types of informational molecules in the DNA continued over time to create receptors within the cells for sensing the evolving needs for life.

This process has resulted in creating receptor sensors for the vary large number of enzymes and hormones that are the informational molecules that control our bodies today.

We will not repeat the details here but you can learn more about these informational carriers in the book: **Hormones Working For You.**

HORMONES
WORKING FOR YOU

BY WALTER PARKS
READ BY STEVEN ROY GRIMSLEY

Conclusions of How Life Began

So what do we conclude about how life began?

Looking at today's life forms and tracing the creation/evolution back, it's pretty obvious that all the organisms living today, even the simplest ones, are far removed from the initial life forms of about four billion years.

Those initial life forms would have been much, much simpler than anything living today. They survived and evolved because they must have had that fundamental property of being able to grow and reproduce and be subject to Darwinian evolution.

The earliest things that actually fit that definition were most likely little strands of nucleic acids. Not DNA yet, which are more sophisticated molecules, but something that could catalyze chemical reactions and that had the blueprint for its own reproduction.

The creation of life was probably not a protracted process using a chemistry that is pretty unlikely but rather **is chemistry** that, when conditions for the recipe get right, it goes, and it goes fairly quickly.

The **recipe for life** is not that complicated in that our bodies are made from a limited number of elements. Most of our mass is carbon, oxygen, hydrogen, sulfur, plus some nitrogen and phosphorous. There are a couple dozen other elements that are in there in trace amounts, but to a first approximation we are just a bag of carbon, oxygen, and hydrogen.

These elements of our bodies came together rather easily because the atmosphere is a bag of carbon, oxygen, and hydrogen as well, even though it is not living.

So the ingredients or recipe for life may be fairly simple, but the question is how does that carbon dioxide in the atmosphere (or methane in an early atmosphere) and water vapor and other sources of hydrogen combine and change from simple, inorganic precursors and become the building blocks of life?

47

Miller and Urey showed us via their experiment that the chemistry was not some improbable chemistry, but a chemistry that is most likely widely distributed throughout our solar system.

Life is a particular form of chemistry in which the chemicals can lead to their own reproduction. Life isn't somehow different from the rest of the planet. Life is something that emerges on a developing planetary surface as part and parcel of the chemistry of that surface.

Life is really part of the fabric of a planet like Earth.

The steps were probably:

1. The basic 20 amino acids of life spontaneously formed in the primordial soup as illustrated in the Miller and Urey experiment;

2. Groups of these created amino acids coalesced to form into several types of more complex molecules including ribosome and RNA;

3. RNA developed into DNA;

4. DNA underwent Darwinian evolution to achieve the first simple one cell life;

5. The DNA of the one cell life grew to include an improved ability to sense its needed water and nutrients in the vicinity of its cell walls; and created receptors for the sensing of these ingredients it needed;

6. The creation of receptors sensitive to the cell's needs began the path to life based on information molecules; thus beginning the escalation of simple cellular life to true life that led to the higher life forms;

7. The creation of receptor molecules continued and evolved to have hundreds and then thousands of various receptor molecules sensitive to factors outside the cell, including enzymes and hormone massagers that evolved with higher life forms.

Let's say a bit more about some of these listed steps.

The early one cell life forms have been termed "trivial replicators" meaning they were not much more capable of progress than non-life crystals as described above. They were capable of processing information only in the passive sense. They relied strictly on basic physics and chemistry of the current environment to support replication. Therefore, only a limited set of cellular products were constructible.

But the later cells which have been termed "non-trivial replicators" developed informational molecules and therefore could harness all of the underlying laws of physics and chemistry to achieve a broader agenda for growth and evolution. They process information in an active sense, enabling the possibility of change in response to the current informational state of the system and its surroundings.

The true origin of life did not occur until there was a transition from the "trivial replicators" to the "non-trivial replicators". This occurred when the cellular molecules that had been chemical and structural molecules took on the third role of informational molecules.

We can deduce how this most likely happened by combining two approaches, extrapolating the current properties of modern organisms backward in time and deducing the step by step evolution of the one cell "trivial replicators" to the "non-trivial replicators".

In modern organisms, RNA is a biochemical mediator, enabling the translation of DNA to protein. RNA is unique in that it can act as both a genetic polymer and a biochemical catalyst.

This has led to the popular "RNA world" hypothesis, where all known life is believed to have descended from an ancestral population of organisms that utilized RNA as their sole major biopolymer prior to the advent of DNA and proteins.

This hypothesis had some conceptual problems but these can be solved by amending the hypothesis to recognizing that RNA was preceded by an alternative genetic polymer such as peptide nucleic acid.

Introducing these pre-RNA molecules allows the evolutionary process to be much easier for the sequence of building blocks from lipids, to peptides and to iron/sulfide complexes because it's easier to synthesize these molecules and the elements for these products would have been much more abundant on the pre-life Earth.

This sequence approach can construct any possible peptide composed of the amino acids with ribosome acting as the supervisory molecule.

This system sequence would then have undergone further evolution to arrive at the DNA/protein world we observe today.

The key to the varied array of molecules is the enormous number of possible sequences that can be composed from the natural set of the 20 amino acids found in proteins. And remember that the Miller and Urey experiment showed how the amino acids could spontaneously form.

Following all of this we recognize that the key to life is the ability for the cells to receive instructions telling them what to do.

The signals providing these instructions are mostly chemical in nature.

Simple cells have sensors that detect nutrients and help the cells to seek food sources via chemical attractions.

Multi-cellular organisms have developed more complex signals including growth factors, hormones, neurotransmitters, and extracellular matrix components. These substances can exert their effects (attractions) both locally and over long distances.

Neurotransmitters are a class of short-range signaling molecules that travel across the tiny spaces between adjacent neurons or between neurons and muscle cells. Other signaling molecules must move much farther to reach their targets. One example is follicle-stimulating hormone, which travels from the mammalian brain to the ovary, where it triggers egg release.

Some cells also respond to mechanical stimuli. For example, sensory cells in the skin respond to the pressure of touch, whereas similar cells in the ear react to the movement of sound waves. In addition, specialized cells in the human vascular system detect changes in blood pressure that the body uses to maintain a consistent cardiac load.

Cells have proteins receptors that bind to the signaling molecules and initiate a physiological response. Different receptors are specific for different molecules. Dopamine receptors bind dopamine, insulin receptors bind insulin, and nerve growth factor receptors bind nerve growth factor, and so on. There are literally hundreds of receptor types in cells and various cell types have varying populations of receptors.

Receptors can also respond directly to light or pressure, which makes cells sensitive to events in their surroundings and to the atmosphere.

Receptors are generally trans-membrane proteins, which bind to signaling molecules outside the cell and subsequently transmit the signal through a sequence of molecular switches to internal signaling pathways.

There are three major classes of membrane receptors that transform external signals to internal signals: via protein action, via ion channel opening, and by enzyme activation.

Because membrane receptors interact with both extracellular signals and molecules within the cell, they permit signaling molecules to affect cell function without actually having to enter the cell. This is important because most signaling molecules are either too big or too charged to cross a cell's plasma membrane.

Not all receptors exist on the exterior of the cell. Some exist deep inside the cell and some even in the nucleus. These receptors typically bind to molecules that can pass through the plasma membrane, such as gases like nitrous oxide and steroid hormones like estrogen.

Once a receptor protein receives a signal, it undergoes a conformational change, which in turn launches a series of biochemical reactions within the cell. These intracellular signaling pathways typically produce multiple intracellular signals for every one receptor that is bound.

At any one time a cell is receiving and responding to numerous signals, and multiple signal transduction pathways are operating in its cytoplasm.

This complexity slowly evolved over long periods of Darwinian time producing the amazing array of living creatures now on planet Earth.

But was it really Darwinian evolution?

Chapter 3
Creation, Evolution, or Divine Guidance

"It is not the strongest of the species that survive, nor the most intelligent, but the one most responsive to change."

Sir Francis Darwin
(1848 - 1925)

We, the authors, grew up in the Christian world of the Deep South. We were taught the stories and lessons from the Bible. God created all life on earth. He created Adam and Eve as the first humans.

But science has proven that man is related to lower life forms.

Ancient fossils gave the first clues when fossils of Neanderthals and other species of pre-humans were found.

But the real proof is in our DNA. We only have to follow the DNA to understand our origins.

When we look at the enormous amount of data on the origins of life and the origins of man, we see that the answers are complex. Did God really create life and man as stated in the Bible? Or did we come into being via evolution? Or did God use evolution as His tool to create life and man?

Let's explore the possibilities.

Humans have approximately 20,000 to 25,000 genes and share 99% of their DNA with the now extinct Neanderthal and 95 to 99% of their DNA with their closest living evolutionary relative, the chimpanzees.

DNA is God's, or if you are a non-believer, Mother Nature's primary tool for creating Man and all life forms on planet earth.

After creating the first Hominoid about 6 million years ago, a minor change was made in its DNA to create the next species. This process continued until about 2 million years ago when the genus Homo was established.

The process then continued until about 200,000 years ago when Homo sapiens came to be and Cain was born. The process is still continuing today.

The genetic difference between individual humans today averages a minuscule 0.1% (.001). **It is only 1.2% between humans and the chimpanzee.**

The DNA difference between us and gorillas is about 1.6%. Most importantly, chimpanzees, bonobos, and humans all show this same amount of difference from gorillas.

A difference of 3.1% distinguishes us "African apes" from the Asian great ape, the orangutan.

All of the great apes and humans differ from the rhesus monkeys, for example, by about 7% in their DNA.

The DNA evidence shows that our human creation/evolutionary tree is embedded within the great apes tree. In the scientific classifications we are classified as a great ape.

Don't get upset; it's just the name of a classification.

The fossil evidence supports this DNA evidence, or should I say that this DNA evidence supports the fossil evidence.

Let's summarize.

Due to billions of years of creation/evolution, **humans share genes with all living organisms, including plants.** The percentage of genes or DNA that organisms share records their similarities. We share more genes with organisms that are more closely related to us.

We have already discussed the very high percentages of DNA that we share with the apes. But we also share high percentages of our DNA with all living creatures. We share 90% with cats, 80% with cows, 75% with mice, 60% with the fruit fly, and 50% with the banana.

Yes, the banana!

Animal and plant life share so much ancient DNA coding because animals and plants had the same ancestors way back and did not diverge until approximately 1.5 billion years ago.

Humans belong to the biological group known as Primates, and are classified with the great apes, one of the major groups of the primate creation/evolutionary tree. Besides similarities in anatomy and behavior, our close biological kinship with other primate species is indicated by DNA evidence. It confirms that our closest living biological relatives are chimpanzees and bonobos, with whom we share many traits.

But we were not created nor did we evolve directly from any primates living today.

DNA shows that our species and chimpanzees diverged from a common ancestor species that lived between 8 and 6 million years ago. The last common ancestor of monkeys and apes lived about 25 million years ago.

You can get more information on this creation/evolution from the book: **Cain's Wife, Lilith's Daughter.**

CAIN'S WIFE
LILITH'S DAUGHTER

BY WALTER PARKS
READ BY ALEX HYDE-WHITE

So it is clear that we were either created, or evolved, from that same "seed" of life previously described.

The DNA proves this beyond any doubt.

The question is did we evolve by happenstance like the atheists believe or did God use evolution as His tool to create us.

We know the Biblical story of Adam and Eve is an analogous summary of how we were created; in light of the scientific evidence we know that the Biblical story cannot be literal. We also know that the Bible was significantly rewritten in the late BC and early AD era.

The scientific evidence has continued to grow, especially over the recent past years, and has caused many who blindly accepted the Biblical story to begin to question its **literal** validity.

Most people have accepted evolution.

But when our research showed that evolution does not provide a definitive answer as to how simple life first developed the "informational genes" previously described, it caused us authors to try to grab this handle to return to the Biblical teachings of our youth.

If you do not believe in God it is easy to believe in evolution.

If you believe in God and especially are a product of early Christian teachings, it's harder to accept.

So the answer to the basic question of the origin of man depends on the answer to the question: Is there a God?

Chapter 4
Is There a God

"If God did not exist, it would be necessary to invent him."

Voltaire
(1694 – 1778)

"Science without religion is lame; religion without science is blind."

Albert Einstein

Three major reasons can be argued for there being a God:

1) Most of us have been taught that there is a God and the concept is a basic part of our culture and may even be in our genetic DNA.

2) The elegant structure and operation of the universe and life essentially demands the need for a creator/designer.

3) The recently learned fact that life requires informational genes to direct DNA so that the cells know what to do and how to work requires intelligence input that, so far, cannot be explained via evolution.

Because of these arguments/reasons we see that almost everyone believes that there is a God.

In a recent survey of 230 countries 84% identify with a religious group:

Christians	32% of which 16% are Catholic.
Muslims	23%
Hindus	15%
Buddhists	7%
Jews	0.2%
Other	6% including African and Native American Traditional religions

16% state that they do not believe in any religion but many of these say that they do believe in some sort of God or universal spirit.

The major argument for there not being a God is the lack of hard evidence. We can only **deduce evidence** for there being a God.

If there is a God who and what is He?

Most all of us have been taught that He created the universe and the earth and all life including us.

Most believe that He is a Personal God to whom we can pray and expect results.

Many believe that He is like us; that we were made in His image.

Many believe that He is fair, but in an "eye for eye" context.

Many believe that he is vengeful but can be forgiving.

Most of us **believe what we have been taught to believe**, even though we understand that the stories we have been taught were originated by men that lived a very long time ago when we did not understand life and the universe as well as we do today. Many of us also know that the stories have been modified down through time, and especially so in the late BC and early AD eras.

Looking back on these stories and their modifications and significant changes we clearly see that there is no real factual basis for the stories. But they have been instilled in our cultures and perhaps even in our genetic memories.

We cannot however, rely on any of these teaching as facts.

So how do we answer the question: Is there a God?

There are no hard, provable facts on either side of the answer.

It has been said that Man invented God to answer the unanswerable. Then down through the ages, as we learned more and more about life on Earth and more knowledge of the universe, we have changed, and lessened, the role of God and He has become less in our daily lives.

Earlier lives were dominated by serving the perceived needs to please God. Few lives are so dominated by service to God in most of today's societies.

So what can we conclude?

We authors decided that, at this point in our narrative we should keep an open mind and continue to search for hard clues for the existence or non-existence of God. We recognize however that we have been too programmed by the teachings of our youth to be completely opened minded; we cannot help but be biased towards our teachings that there is a God.

We recognize that down through the ages Man has "used" God as the answer to the unanswerable. So we will continue to "use" the existence of God to answer our continuing unanswerables.

But we will continue to seek answers to the unanswerable. We just might be able to see that God continues to exist and that His works of wonder cannot be completely answered by the evolutionary forces of the laws of physics.

We will revisit the question of the existence of God after we have explored the possibility of the soul that may provide us with an afterlife. We will also continue to search for God in our explorations of our elegant universe.

Chapter 5
Do we Have a Soul that Survives Death

We must no more ask whether the soul and body are one than ask whether the wax and the figure impressed on it are one.

Aristotle
384 BC – 322 BC

We humans know that we're going to die. We hope that we will have some form of afterlife. But we know that our bodies decay. Therefore in order to have an afterlife there must be some part of us that survives death. We have come to believe in the concept of the soul.

We have never understood what the soul really could be but we know that if we survived death that we must have something like the soul.

If we survive death we assume that we could only do so if we have a soul. We therefore decided to explore and **search for a scientific possibility for the soul.**

We know that our five senses cannot detect the soul. We therefore know that if we have a soul we would have to find it in some manner beyond our five senses. We thought about how Einstein used his imagination to develop the theory of relativity. He could not detect the parameters of relativity with his five senses. So he imagined things beyond his five senses and came out with his famous equation $E = MC$ squared where E stands for energy, M stands for matter and C is the speed of light.

Let's use his equation to explore for the soul.

Our Body An Energy Matrix of Matter
Einstein taught us that there is a definitive relationship between matter and energy. Take another look at his famous equation: $E=MC^2$ and take particular note of the relationship between mass, i.e. matter (M) and energy (E).

Our flesh and bones are matter, but matter is a form of energy. Therefore our bodies are energy matrices of matter.

Now let's take a look at that matter. We will use the model we were taught in high school.

The matter of our bodies is made of atoms. The atoms have a center nucleus of protons and neutrons. Electrons orbit about the center of the atom. There is a tremendous distance between the nucleus and the orbiting electrons, that is, compared to the size of the nucleus and the electrons. **So our bodies are mostly made of space**.

The matter within this mostly space matrix is structured and organized by our DNA which is just a coded form of matter. It carries the code to structure our bodies in our mother's womb and instructs for our growth. It provides the operating instructions for day to day operations of our bodies.

Now step back and visualize our bodies. Imagine a body magnified to the size of a galaxy. Really? Stay with us.

We could then use a magnifying glass to see the many specks of atom nucleuses, but we would mostly see space. We see a picture not unlike what the Hubble telescope sees when it looks out at a galaxy in space.

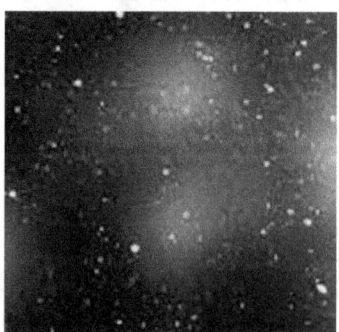

Isn't imagination wonderful? It can really help us better understand concepts beyond our 5 senses. Einstein was a genius at it.

Now let's go a bit further in our imagination. Visualize the matter "decaying" to pure energy according to Einstein's famous equation shown above. You may prefer the word transforming rather than decaying.

Is the body still the same body but now in an energy matrix of energy rather than an energy matrix of matter? Stay with us.

Now the energy matrix may not need all that space that the atoms needed. In fact, it does not need any space. It does not need to be a specific size as are our bodies. This new version of the body can be as large as the galaxy size we imagined or as small as we are or maybe even smaller. Actually it has no size at all, as we envision size.

Maybe we should mention that space is one of the eleven dimensions of universes just as are the spatial dimensions of up-down, left-right and back and forward; and time. But that's another story for another more complex scientific book.

This energy version of our body cannot be detected with our 5 senses. But it can be detected with electromagnetic sensors because this energy matrix is, that is it produces an electromagnetic field.

Truth be known, the body is always a flesh and bone matrix of matter and energy and is therefore also an electromagnetic field.

So our bodies have always had a physical being as we detect with our 5 senses, and an electromagnetic being as can be detected with an electromagnetic sensor.

Recognizing these facts arms us with the ability to take our imaginations and speculations to another, very much more amazing level.

Our Soul: An Energy Matrix of Pure Energy

How should we characterize the pure energy that is the electromagnetic version or signature of our body? We can't see it. We can't touch it. We cannot detect it with our 5 senses.

What should we call it? The often used word spirit probably fits as good as any.

Could this spirit be our soul? Could it survive when our physical bodies have ceased to be?

Could the pure energy form of our DNA continue to have the ability to form our body and to "operate" our spiritual body?

If provided with external energy could it follow the code in its pure energy form of our DNA and re-materialize some energy form, spiritual form of our body?

Can we visualize this being our soul?

Think about it.

Visualizing the Soul

OK, let's visualize how this could be.

While we are alive our DNA instructs our bodies to use ingredients from the foods that we eat to manufacture proteins, hormones and other materials to grow, repair and operate our bodies.

Now after we die we can access some of this DNA to rebuild (clone) our body as was accomplished in the famous Dolly the sheep project. We have not yet done this with a human, as for as the public knows, because of ethics problems. But we could clone a human just as we did Dolly.

So DNA can continue to code for the structure and operation of our body even after death.

Now the DNA from our deceased body is in its material form just as it was in our bodies when we were alive.

Ok, but the DNA of our dead body can still be active for a while. And while it is still matter, it is also still a form of energy: $E=MC^2$.

Now let's suppose that the energy form of our DNA remains even as our material, matter form, decays and returns to dust. If it could remain then it could be the surviving form of the body. **We are our DNA, the material form in life and the energy form after death.**

This surviving "spirit" or energy form of our DNA could be no more capable of doing anything or thinking anything than isolated DNA cut from our living bodies.

But what if we could stimulate this energy form of our DNA with nutrients like it is stimulated with nutrients from food when it was in our living body? What if this energy form of our DNA could be stimulated and able to use pure energy provided from an external source?

Remember your high school physics: energy is energy; it can be neither created nor destroyed.

If we can trigger the energy form of the DNA to be receptive to energy that it can draw from its environment, then it could rebuild the energy form of the body. It would not need the years of time that the matter form of the body needed to assemble all the matter based nutrients required for the slow growth of the human body. Energy conversion need not require such time, if any time at all.

Forget the limiting ability of your 5 senses; free yourself to imagine.

You already know that, as Einstein proved to us, there are worlds beyond our 5 senses. Go into that unlimited world.

Now "visualize" the pure energy form of DNA as described above.

Take another step into that unlimited world, "visualize" the pure energy form of the DNA sucking up enough energy from the surroundings to almost instantaneously reform the entire energy form of the body.

The completeness of the reformation will depend on the amount of energy available. We may be able to detect this "electromagnetic entity" as a small orb with our electromagnetic sensors.

Orb Claimed to be a Soul in a Cemetery

Or we may be able to detect a human size orb, depending on the energy available.

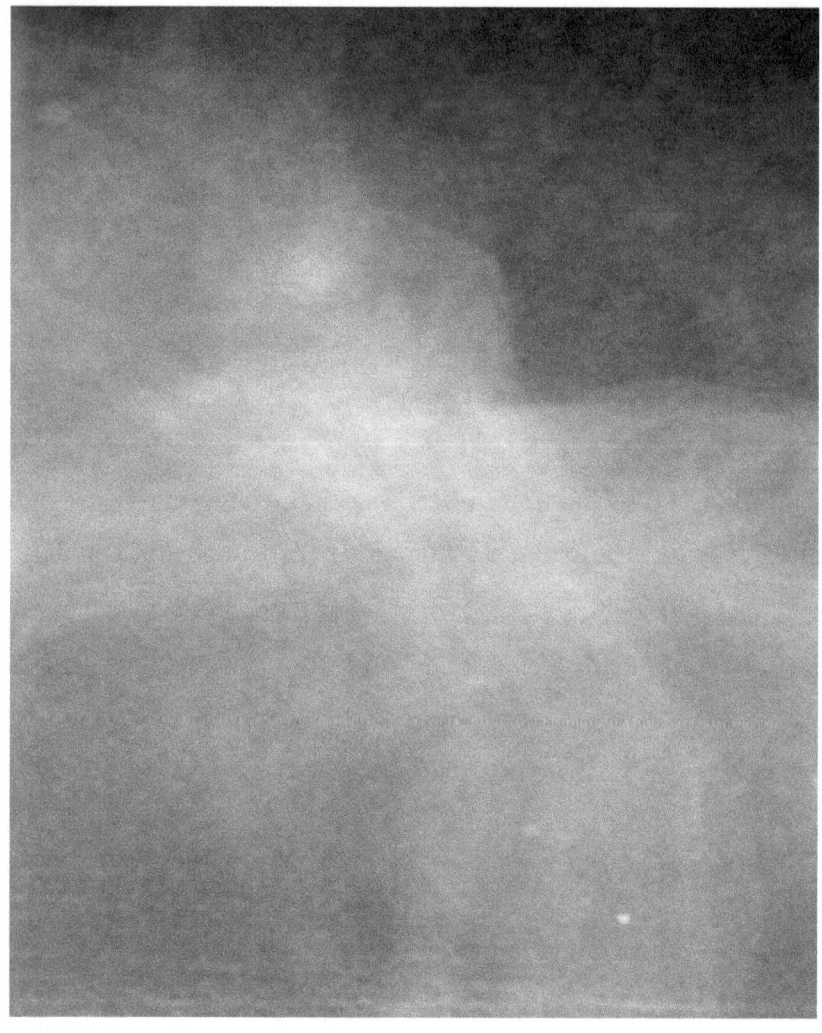

And if there is enough energy, some people may be able to "see" the reformed, functioning, energy form of the body. Many people have claimed that they have not only seen these "spirits" but claim to also have communicated with them.

Amy was able to do this as described in the fiction book: **Paranormal Portal to a Parallel Universe.**

PARANORMAL
PORTAL TO A
PARALLEL UNIVERSE

BY WALTER PARKS
READ BY DAVID GILMORE

Activating the Soul

If you are still with us, that is if you have kept your imagination open, let's proceed.

Now so far we have described a possible scientific explanation of the soul without violating any laws of physics. But how long could that pure energy form of our DNA survive? And what could activate, that is cause the DNA to draw energy from the surroundings to form the pure energy version of our body?

Paranormal investigators and psychics have speculated that someone dying in horrible or at least unusual circumstances could cause the spiritual body to return to perhaps right a great wrong. Can we speculate how this could happen?

We can speculate that the spirit may "hang around" where they experienced the trauma. Let's assume that scenario for the moment.

We can also speculate that this pure energy form of each of us survives, at least generally, forever. We can not think of any reason for this pure energy to cease to exist. Remember you can neither create nor destroy energy. But it could, perhaps eventually, be absorbed with other energy. But that's another story in another, more religious book.

So let's assume that the spirit is hanging around the site of the trauma and something happens so that the pure energy form of the DNA is triggered. In order for anything to trigger the DNA, the DNA would have to be capable of sensing the trigger.

We find it hard to believe that DNA could sense a trigger although it does sense the presence of certain RNA and hormones in the living body. We think it may be better to assume that the sensor would be the spiritual being itself. We perhaps should therefore assume that the spirit can continuously suck enough energy from its surroundings to continuously have an existence. Then when it senses the trigger it materializes more completely, but in energy form.

When the spirit sucks enough energy from its environment to materialize, the surrounding will get colder. Almost all of those experiencing a paranormal event or a ghost experience have said that their presence made the area cold.

We'll settle for this assumption for now. We will continue to think about this and speculate farther. We hope that you will too. We hope that our speculations about the soul are not too "New Age" for you to consider.

We will publish more on the soul after we collect more evidence from research of paranormal events.

History of the Soul

We developed this energy matrix version of the soul in our efforts to try to find a "somewhat" rational, scientific explanation for the soul. Let's compare our concept with the historical concepts of the soul.

Mankind has believed in the concept of a soul for ages. Various philosophies and religions have many differing views about the soul. It is generally viewed as the "immaterial aspect or essence of a human being that confers individuality and humanity." Some consider it synonymous with the mind or the self.

Generally theology defines the soul as that part of the individual which partakes of divinity. Many religions considered it to be that part of us that survive the death of the body.

There is evidence that even prehistoric peoples recognized some incorporeal principle of human life or existence corresponding to the soul.

Although beliefs in the soul are widespread and longstanding, various religions and philosophers have developed a variety of theories as to its nature, its relationship to the body, and its origin and mortality.

Some of the early ancient Egyptians and Chinese conceived of a soul which survives death. This concept led some to ancestor worship.

The early Hebrews apparently had a concept of the soul but did not separate it from the body, although later Jewish writers developed the idea of the soul in more details but as they continued to write down through the ages, the different writers created a variety of concepts of differing views.

Christian concepts of a body-soul dichotomy originated with the ancient Greeks and were introduced into Christian theology at an early date by St. Gregory of Nyssa and by St. Augustine. The Ancient Greek concepts of the soul varied considerably according to the particular era and philosophical school.

The Epicureans considered the soul to be made up of atoms like the rest of the body. For the Platonists, the soul was an immaterial and incorporeal substance, akin to the gods yet part of the world of change and becoming. Aristotle's conception of the soul was obscure, though he did state that it was a form inseparable from the body.

In Christian theology St. Augustine spoke of the soul as a "rider" on the body, making clear the split between the material and the immaterial, with the soul representing the "true" person.

From the Middle Ages onward, the existence and nature of the soul and its relationship to the body continued to be disputed in Western philosophy. To René Descartes, man was a union of the body and the soul, each a distinct substance acting on the other; the soul was equivalent to the mind. To Benedict de Spinoza, body and soul formed two aspects of a single reality.

Immanuel Kant concluded that the soul was not demonstrable through reason, although the mind inevitably must reach the conclusion that the soul exists because such a conclusion was necessary for the development of ethics and religion.

Just as there have been different concepts of the relation of the soul to the body, there have been numerous ideas about when the soul comes into existence and when and if it dies. Ancient Greek beliefs were varied and evolved over time. Pythagoras held that the soul was of divine origin and existed before and after death. Plato and Socrates also accepted the immortality of the soul, while Aristotle considered only part of the soul, the intellect, to have that quality. Epicurus believed that both body and soul ended at death.

The early Christian philosophers adopted the Greek concept of the soul's immortality and thought of the soul as being created by God and infused into the body at conception.

In Hinduism the cycle of death and rebirth is considered eternal according to some Hindus, but others say it persists only until the soul has attained karmic perfection after which it merges with Brahman.

The Muslim concept, like the Christian, holds that the soul comes into existence at the same time as the body; thereafter, it has a life of its own, its union with the body being a temporary condition.

We believe our pseudo scientific concept is more believable than any of these historical beliefs.

Where does the Soul Reside

It is accepted that, if a soul exists it resides in the body while we are alive. But where does it reside after death?

Some people, principally psychics believe that the soul hangs around the place of death or trauma and can be communicated with at those places. The book **Paranormal Portal to a Parallel Universe** describes a fictional account of such a paranormal encounter.

But how long do they "hang around" at those places? Do they always reside there or eventually go elsewhere? And what about souls that have no reason to hang around at specific places, where do they reside?

Is it possible that they reside in a parallel universe that is within inches of our universe? Let's take a look.

We are learning that parallel universes really do exist. We describe the evidence for the existence of Parallel Universes in Chapter 11.

NASA has installed the Alpha Magnetic Spectrometer II (AMS-2) to help them look for proof for the existence of parallel universes.

It is believed by some that parallel universes are made of dark matter for which AMS-2 is designed to find.

AMS-2 is attached to the International Space Station (ISS) which orbits the earth at an altitude of 210 miles.

This is the first time a magnetic spectrometer has been launched into space. It will search the stars and galaxies millions of light-years beyond our Milky Way galaxy. The elements in the universe that we see are composed of matter. We now know that the universe contains additional properties called dark matter and dark energy, which are described in some detail in a later chapter.

Further study of these properties is expected to prove that parallel universes do exist.

The particles that will be detected by the AMS-2's 7.5 ton device cannot be detected on earth because of earth's atmosphere and electromagnetic shield.

It is expected that AMS-2 will provide data that will prove the existence of parallel universes.

It is also believed by some that particles discovered by AMS-2 could verify the M-Theory and help us to explain the Big bang without the need for a singularity. Most scientists have always had a hard time in believing in the singularity where all equations of physics become invalid.

It is imagined that parallel universes may be as close to each other as a few meters or even closer. The only observed interactions across such tiny gaps are the gravitational pull to each other.

AMS-2 will provide the first time that charged particles can be studied in the cold vacuum of space away from the distorting influence of Earth's presence.

AMS-2 is expected to provide data that will create a true revolution in our physics in a manner similar to the revolution Einstein's theory of relativity caused to Newtonian physics.

So, assuming that AMS-2 will prove the existence of parallel universes, which many scientists already believe, let's speculate that our souls may reside in such a parallel universe.

For those of us who want to believe in existence after death; for those of us who want to believe in a soul, we need to consider that our souls may reside in an almost infinitely close parallel universe.

Will we soon learn, with the help of AMS-2, enough about parallel universes to justify a more detailed study of our souls residing there?

Will there come a "day of reckoning" or an "apocalypse" when all of our souls will go there and be together?

We know that this is heavy speculation.

But having our souls reside in a parallel universe could help resolve many of the unexplained paranormal encounters.

And it could scientifically explain how a soul could exist.

We'll revisit this after AMS-2 proves the existence of parallel universes and we have researched several more paranormal encounters.

Meanwhile think about it.

You can read more about the soul in the book: **Finding the Soul, Surviving Death**.

Let us know your thoughts at info@UnknownTruths.com

Chapter 6
Can We Achieve Immortality on Earth

I don't want to achieve immortality through my work. I want to achieve it through not dying.

Woody Allen

And out of the ground the Lord God made to spring up every tree that is pleasant to the sight and good for food. <u>The tree of life</u> was in the midst of the garden, and the tree of the knowledge of good and evil.

Genesis 2:9

Then the Lord God said, "Behold, the man has become like one of us in knowing good and evil. Now, lest he reach out his hand and take also of the tree of life and eat, and live forever—" therefore the Lord God sent him out from the garden of Eden to work the ground from which he was taken. He drove out the man, and at the east of the garden of Eden he placed the cherubim and a flaming sword that turned every way to guard the way to the tree of life.

Genesis 3:22-24

Does this tell us that there was a Tree of Life and that if we eat of it we will become immortal?

Let's explore.

In the Vedic lore of ancient India, it is said that:

"In the land of the fathers.... In the shining Paradise" that the Kings sat beneath a tree and drank soma while listening to flutes being played.

<u>They drank soma because it gave them immortality.</u>

The "Shining Paradise" is believed to refer to the land from which the early Vedic peoples migrated after the great flood. The Shinning Paradise was a great civilization that existed prior to the flood. Plato said it was Atlantis.

You can read more details about Atlantis in the book: <u>Atlantis The Eyewitnesses</u>.

Atlantis or not – it was a civilization that existed prior to, and was destroyed by, the flood.

The ancient Rigveda of India devotes its entire ninth book to the praises of soma. It says that soma juice was the nectar of the gods; the **elixir that gave them immortality**. But it also says that mortals could use soma juice.

Soma was so important, that in the later legends of ancient India, it was elevated to a god:

This Soma is a god; he cures

The sharpest ills that man endures

He heals the sick, the sad he cheers,

He nerves the weak, dispels their fears.

The faint with martial ardor fires,

With lofty thought the bard inspires.

The soul from earth to heaven he lifts,

So great and wondrous are his gifts.

Men feel the god within their veins,

And cry in loud exulting strains.

We've quaffed the Soma bright

And are immortal grown.

We've entered into light

And all the gods have known,

What mortal now can harm,

Or foeman vex us more?

Through thee beyond alarm,

Immortal god, we soar.

Note especially:

"We've quaffed the Soma bright

And are immortal grown."

We drank Soma and became immortal! Is this just a wish? Could it be documentation of what they actually did?

You can read more details about ancient immortality in the book <u>Immortal Again.</u>

IMMORTAL AGAIN

ANCIENT SECRETS CAN INCREASE OUR LONGEVITY

WALTER PARKS

Depiction of Ancient God with Soma Purse

Ancient Sumer and Babylon

We have also found references to this Soma plant in the Epic of Gilgamesh, written in the clay tablets of ancient Sumer and Babylon. These tablets note that they were copied from even more ancient texts.

They tell of King Gilgamesh who goes on a mission to seek eternal life. He is told of a plant that can provide him immortality. The ancient document refers to the Soma plant used by the "Kings and Nobles" prior to the flood.

In the ancient clay tablets re-write of several thousand years ago, the story says that the plant is now in deep water, probably an allegory for the flood, which destroyed the plant; made it become extinct.

Gilgamesh, being the hero of the story, of course dives and gets the plant, but it is taken away by the snake. This is the continuation of the allegory in that the snake caused the food and destroyed the life pro-longing Soma plant. The snake and the flood are described in detail in the book about Atlantis referenced above.

An Ancient Drawing of King Gilgamesh

The Bible says that before the flood such a plant grew in the Garden of Eden:

"And the Lord God planted a garden eastward in Eden…."

Genesis 2:8

And we've been taught that that same old serpent induced Eve to eat of the tree of knowledge. And thus God kicked Adam and Eve from Eden:

"…. Lest he put forth his hand, and take also of the tree of life, and eat, and live forever."

Genesis 3:22

So we see many references to this plant, this tree of life, in literature from almost all of the ancient cultures.

The Bible tells us that:

"…. And all the days of Methuselah were nine hundred sixty and nine years …."

Genesis 5:27

Depiction of Methuselah at age 969

And the Bible tells us that others at that time also lived to such old ages, as shown in the table.

Patriarch	Age
Adam	930
Seth	912
Enos	905
Cainan	910
Mahalaleel	895
Jared	962
Enoch	365
Methuselah	969
Lamech	777
Noah	959

Patriarch Ages in the Bible

 Many of the old Biblical Patriarchs lived very long lives before the flood.

King	Reign in Years
Gaur	1200
Gulla-Nidaba-Annapad	960
Pala-Kinatim	900
Nangishlishma	Tablet Damaged
Bahina	Tablet Damaged
Buanum	840
Kalibum	960
Galumum	840
Zukakip	900
Atab	600
Mashda	840
Arurum	720
Etana	Unknown

Reigns of the Sumerian Kings

The clay tablets from ancient Sumer and Babylon say their Kings lived comparable life times. The tablets list their reigns rather than their ages, but we see from the chart that their life times were comparable to the Biblical patriarchs.

Some of us were taught from childhood to believe in the Bible, based on faith alone. But others smile at the absurdity of living to be 969 years and attribute such statements to just another old myth.

But is it a myth? Is there any scientific evidence for any of this?

Yes. Let's take it a step at a time. Let's understand why we age and what we can do to become Immortal Again, like the ancients.

There are a dozen or so mechanisms of aging that have been theorized over time. The authors believes that there are seven basic causes that combine to make us age:

1. Free radicals and other ashes of our metabolism, and environmental toxins build up in our cells and cause the cells to die without being able to reproduce themselves as they would normally do when you are younger.

2. Our endocrine system ceases to secrete sufficient quantities of certain enzymes and hormones to keep up with the cell's battles with the build up of contaminants.

3. Our cells lose their ability to divide, and replace themselves because they use up their allotted number of divisions (reach their Hayflick Limits as explained later).

4. Stress causes secretion of excessive cortisol which does significant damage.

5. Some of us have inherited flawed genes that cause, or allow malfunctions.

6. Deficiencies in our diet limit the materials necessary for the cells to cleanse and repair themselves. Excesses in our diets also adversely affect certain chemical reactions.

7. Lack of exercise causes atrophy of critical muscles that result in chemical imbalances and loss of strength and agility which makes us prone to accidents.

The 7 causes of aging are described in detail in the book **Aging is a Treatable Disease**.

AGING
IS A TREATABLE DISEASE

BY WALTER PARKS
READ BY DOUGLAS R PRATT

Aging cause number 3 above is the key to immortality. Our cells lose their ability to divide, and replace themselves because they use up their allotted number of divisions (reach their Hayflick Limits).

Dr. Hayflick determined that cells need to replace themselves as the ashes of metabolism and other toxins build up the cells over time. Each time they divide to produce daughter cells to replace the aging cells they lose some of their telomeres.

Our chromosomes which carry our DNA are in the center nucleus of our cells. At each end of each of the paired chromosomes there is a telomere. It is analogous to the grommets at the ends of shoelaces. The telomeres keep the chromosomes organized.

Telomeres at Ends of Chromosome

As the cells age and become clogged the cell divides to reproduce its self. In so doing, it loses some of the telomere length.

Dr. Hayflick determined that most normal cells, after the initial rapid growth of the embryo, can divide on average about 50 more times after birth before their telomeres are too short for further divisions.

[It is noted that we interviewed Dr. Hayflick for information contained in <u>Aging is a Treatable Disease</u>. We plan to publish a video of this book when time permits. The book is currently in 3 formats: eBook, Paperback and Audio Book at <u>Amazon</u> and most places where books are sold.]

As our cells age and become less effective, they divide to produce new cells to replace themselves. But the more times that they divide, the less effective the cells become at reproducing themselves. Eventually their telomeres become too short for the cells to reproduce.

Baring poor nutrition and a poor environment, the telomeres of most of our cells may allow replacement for up to the age approaching 120 years.

Some have noted that this age limit is not unlike the limit of life span quoted in the Bible. This limit is now termed "The Hayflick Limit" because Dr. Hayflick was the one that discovered the telomere function.

The Bible tells us of the long lives of the patriarchs but then tells us that after the flood:

"Then the LORD said, 'My Spirit will not contend with man forever, for he is mortal; his days will be a hundred and twenty years."

Genesis 6:3

This tends to support the King Gilgamesh story that Man lived long lives by drinking soma but that the flood destroyed the soma plant, the Tree of Life and therefore life was limited to the natural Hayflick limit.

So to review, the telomeres are non-producing genes at the ends of each chromosome that act like grommets at the ends of shoe laces to prevent the chromosomes from becoming damaged.

When the cells have divided a number of times equivalent to 120 years of life they become too short for further divisions so the cells die without being able to replace themselves. So our natural lives are limited to 120 years unless we can do something about the telomeres.

It is interesting to note that Dolly, the famous cloned sheep died earlier than expected. That is because the cells from which Dolly was cloned were already several years old. They reached their Hayflick limit earlier than a naturally born sheep with young genes would have.

So what can we do to "reset" our telomere death clock?

Can we become Immortal Again like the Biblical Patriarchs and Sumerian Kings of yesterday were before Noah's Flood destroyed their Soma?

Can we combine the lore of yesterday with the new DNA and Human Genome knowledge of today to recreate The Tree of Life and become Immortal Again?

Yes, it appears that we can.

We see that King Gilgamesh, the ancients in Sumer and Babylon and Methuselah and the ancient Biblical Patriarchs, according to ancient literature, found the means to live very long lives. They were essentially immortal from the standpoint of aging.

Their ancient knowledge and the means to extend longevity were apparently lost, most likely due to the great flood when almost all of mankind was destroyed.

We are just now beginning to learn what they knew in antiquity. The relatively recent decoding of the Human Genome is helping us to much better understand our genes and how they work.

It is very interesting to apply our new understandings to the ancient technology.

Understanding the double helix and telomeres of our DNA helps us understand what the ancients knew over 10,000 years ago.

Resetting the Death Clock

Tests over the past few years have indicated that the "Hayflick limit" may be extended by the use of an enzyme that causes the "organizing genes" at the ends of the chromosomes (the telomeres) to re-grow. This enzyme is called telomerase.

We learned about telomerase from one of our worst enemies, cancer.

A few years ago, cancer researchers noted that some cancer cells divide and reproduce extremely fast, and many more times than the Hayflick limit. It was learned that these cancer cells secrete an enzyme named telomerase. Telomerase causes the telomeres at the ends of the chromosomes to grow such that their cells can continue to divide.

These cancer cells appear immortal.

The researchers became both excited and concerned; excited that cells could become immortal; concerned that such cells would always become cancerous.

A series of experiments indicated that their concern was, perhaps, unfounded. The experiments consisted of treating various cells, including human cells from the foreskin of circumcised infants with telomerase. It was proved that telomerase could be used without inducing cancer.

Dr. Harley, et al at the Geron Corporation recently used the telomerase enzyme in lab tests to re-grow eroded telomeres in human cells. The cells then divided (growing new cells) many more times than the natural Hayflick limit would normally allow. They made human cells become immortal by the use of telomerase.

[We interviewed Dr. Harley, et al to get information for this book and this will later be included in the video mentioned above.]

The protein structure of telomerase is similar to an enzyme found in a certain plant [the Tree of Life?] A "cousin" of this plant with the exact protein structure of telomerase, may have lived before - and was destroyed by - the flood.

Now after the flood no one can get the telomerase enzyme so our telomeres at the ends of the chromosomes continued to erode as the cells split to form new cells. After a certain number of splits the telomeres become too short to continue to divide and our life span is limited to our current "natural lifespan" of 120 years.

The Bible tells the same story, but in a more poetic manner.

There are at least 3 potential approaches to becoming Immortal Again.

1. Genetically Produce a Soma (Telomerase Plant)

2. Find a way to re-activate, i.e. turn back on, the genes that enabled us to produce telomerase while embryos in the womb

3. Clone bacteria to produce human telomerase in the manner in which we have cloned bacteria to produce insulin for diabetics and for human growth hormone (HGH) for the anti-aging programs.

Genetically Produce Soma (Telomerase Plant)

The protein structure of telomerase has been found to be very similar to an enzyme found in a certain plant. This research work is proprietary so the name of the plant is considered confidential.

A "cousin" of this plant with the exact protein structure of telomerase, is believed to have lived before - and was destroyed by - the flood.

It is proposed that we genetically modify this existing plant to produce the telomerase enzyme. We would then use one of the already developed methods to get the telomerase to all of our cells.

We would then run a ton of tests on animals and then humans to verify the effectiveness and safety.

This approach appears very feasible.

Conventional breeding techniques have been in practice for hundreds of years. These techniques are now giving way to genetic alterations. Genetically altered crops have been planted since 1995.

In 1955 a gene from a strawberry plant was inserted into a mustard plant so it would produce a chemical lure to get predator insects to come to it to protect it from mites and other insects harmful to the plant. This genetic modification delivers the plant's defense against insects in its seed rather than having the farmer have to use a spray. This is because the genetic modification becomes part of the plant's DNA; the modified plant becomes a new plant; its life is perpetuated in its seeds.

The study team reports that the genetically engineered plant was able to attract predatory mites (a small relative of spiders) that prey on plant-eating spider mites. The plant needed little genetic modification to introduce the chemical lure. The machinery to produce this "alien protein" is available in all plants. You just have to tap into the existing pathways [in plant cells], and that may be done by the introduction of just one gene.

This research is expected to allow humans to harness plant-bug interactions to improve essentially all future crop yields.

During the last five years, several large companies, including Monsanto and Syngenta have altered our food supply with genetic modifications. Genes from bacteria, viruses, foreign plants and animals have been inserted into corn, soybeans, potatoes, tomatoes, squash, and papayas. These corporations plan to "genetically engineer" almost 100% of our food within a decade.

Already about 40% of the soybeans, 20% of the corn, and a percentage of the potatoes grown in the U.S. and about 50% of the canola (rapeseed) plants in Canada have been genetically altered.

More than 60% of the packaged food items in US grocery stores contain genetically altered ingredients.

Scientists have known how to get plants to resist pests and herbicides for years. It was demonstrated in 1995. Genes from other plants and animals can be inserted to allow the altered plants to produce their own insecticides.

What is new in 2012 is the ability to induce plants to create new products by tinkering with the plants' own synthetic pathways.

Sarah O'Connor of the Department of Chemistry at MIT has studied periwinkle for several years because it produces a variety of compounds that may be used to treat cancers and hypertension. She has now produced new compounds by manipulating the complex biosynthetic pathways of the periwinkle plant. This sort of manipulation offers a new way to tweak potential drugs to make them more effective and less toxic.

Earlier, in 2007, several plants were genetically modified to remove toxic compounds from the environment.

One of the research groups used small plants related to cabbage and mustard to clean up soil contaminated with cyclonite, or RDX. The widely used RDX explosive is highly toxic and carcinogenic.

Another research team modified a poplar tree to soak up a host of cancer-causing compounds from soil, groundwater, and air. The genetically altered plants break the contaminants down into harmless compounds in a process called phytoremediation.

"It is our hope that by developing trees that can remove carcinogens from the water and air in a fast and economical way, people will be more likely to use [the land] than abandon the property as too expensive to clean up," said Sharon Doty, of the University of Washington.

It should also be noted that we have also placed the genes from a pig into a tomato to provide the tomato with a longer in-store shelf life before it begins to soften and rot.

With all of this research and actual applications, especially in the food crop industry, it will just be a matter of time before the genetic alteration of plants to produce telomerase is accomplished.

It will necessarily be led by the development of an effective and safe method to insert the telomerase into the cells of our bodies.

Develop Method to Insert Telomerase in Our Cells

Various techniques have already been used to alter genes by inserting genetic materials from other plants and animals into the gene to be altered.

For plants, the most common vector is a bacterium called Agrobacterium tumefacsiens. The bacterium has the ability to incorporate DNA into plant genetic material. This is a possible way to inject the altered gene for the production of telomerase into the plant to be used to produce telomerase.

Viruses also are good gene vectors for both plants and animals. The viral infection process involves the integration of viral genetic material into the host genome. Viruses can thus be engineered to deliver DNA, which has been introduced into the viral genome prior to infection.

The altered virus then "infects" the plant with the genes to produce telomerase.

Gene insertion can also be achieved mechanically by microinjection of the gene into a cell, or ballistically, by shooting gold beads coated with the gene of interest into the cell. This is not controllable enough for the telomerase treatment in developed humans. It could be used at the embryo stage.

One or more of the more recent developments in gene therapy research will most likely be the methods first used to inject telomerase in developed humans.

Good results have already been experienced in a process called adoptive cell transfer. Researchers drew a small sample of blood that contained normal lymphocytes from individual patients and infected the cells with a retrovirus in the laboratory. The retrovirus acts like a homing pigeon to deliver genes that encode specific proteins, called T cell receptors, into cells. These make more receptor proteins to coat the outer surface of the lymphocytes. The T cell receptors act as homing devices in that they recognize and bind to certain molecules found on the surface of tumor cells. The T cell receptors then activate the lymphocytes to destroy the cancer cells.

In this test, the newly engineered lymphocytes were infused into 17 patients with advanced metastatic melanoma. One month after receiving gene therapy, all patients in the last two groups still had 9 percent to 56 percent of their T cell receptors-expressing lymphocytes. There were no toxic side effects attributed to the genetically modified cells in any patient.

In 2005 Gene Therapy successfully cured deafness in guinea pigs. Each animal had been deafened by destruction of the hair cells in the cochlea that translate sound vibrations into nerve signals. A gene, called Atoh1, which stimulates the hair cells' growth, was delivered to the cochlea by an adenovirus. The genes triggered re-growth of the hair cells and many of the animals regained up to 80% of their original hearing thresholds. This study, which many pave the way to human trials of the gene, is the first to show that gene therapy can repair deafness in animals.

In early 2003 a research team at the University of California, Los Angeles was able to get genes into the brain using liposomes coated in a polymer call polyethylene glycol (PEG). The transfer of genes into the brain is a significant achievement because viral vectors are too big to get across the "blood-brain barrier." This method has potential for treating Parkinson's disease.

Sickle Cell Disease was successfully treated in mice in 2002.

This research and clinical studies are rapidly producing a body of work that can lead to the successful injection of telomerase in perhaps a shorter time than previously thought

Re-activate Our Telomerase Genes

Perhaps this is the most promising approach to get enough telomerase to keep the telomeres from eroding. This involves re-activating the genes that produced telomerase when we are embryos in our mother's womb. All of us carry these genes when we are embryos but cease to produce telomerase after birth.

We need to find a way to re-active, i.e. turn these genes back on especially so as we get older.

The decoding of the human genome in 2003 provided us with the fundamental understanding of what our genes are and what they do. We, the authors are now working with a group to implement this technique.

We believe that this is the best of all methods for extending our life spans and maybe even for achieving immortality.

We already know, at least to some extent, how genes are turned on and off. Turning genes on and off is a major activity of all living cells. Almost 10 percent of the genes in the human genome are switches for turning genes off or on.

The University of Michigan at Ann Arbor first explored the mechanisms for turning genes on and off 55 years ago. They studied gene regulation in bacteria and discovered that sugars in the food supply turn on the genes required for their own digestion. In addition, when bacteria are transferred from a medium containing the sugar lactose to a medium without lactose, the bacteria turn off their lactose-metabolizing genes.

The mechanism for this switch is based on a regulatory protein called the lac repressor.

In the 1970s and 1980s, scientists discovered that gene regulation in mammals also uses the mechanism of protein recognition of short DNA sequences. These short regulatory sequences are called enhancers. For example, the hormone testosterone binds a receptor protein that recognizes a 15 base-pair DNA sequence. As a result, genes that contain this sequence can be activated by testosterone.

Estrogen, in contrast, regulates a different set of genes that have their own distinct sequence. Researchers can exploit enhancers in experiments by fusing a known enhancer to a gene that they want to regulate.

As an example, one might fuse the estrogen enhancer to the hemoglobin gene and insert the construct into the chromosome of a mouse. When the resulting mouse is treated with estrogen, the hemoglobin gene will be turned on.

Other regulators respond to changing environmental signals, such as the amount of protein or trace metals in the diet. Scientists are using the knowledge about gene regulation combined with some newly developed cloning tools to investigate the physiological effects of switching specific genes on or off in aging adults.

A California biotech company has a technology that can turn any gene on or off. The technology allows a scientist to genetically engineer a protein with what is called a zinc finger. Heart not pumping hard enough for lack of good blood vessels? Turn on the blood vessel-growing gene. Want to stop a patient from getting fatter? Turn off the gene that makes fat cells. [I am anxiously awaiting this one!]

But biology is not yet up to speed with zinc finger technology. But it is comforting to believe that someday we will be able to re-grow arteries to replace the ones we have clogged with our undisciplined lifestyle; and to finally get rid of our excess weight.

Zinc fingers occur naturally inside the nucleus of all organisms, where they bind to DNA to turn genes on or off. They are the most common vehicle that genes use for alternating what protein they do or do not produce.

It is just a matter of time before all this new technology and findings are used to develop a telomerase treatment.

It is just a matter of time before we can become Immortal Again.

Time Magazine boldly stated on their cover of February 21, 2011:

"2045: The Year Man Becomes Immortal."

Can you last that long?

Maybe you had better start some of the more readily available anti-aging programs to insure that you will be around when the immortal treatments become available.

The previously mentioned book: <u>**Aging is a Treatable Disease**</u> lists several programs

But we the authors believe that telomerase treatment will be available much sooner. Human tests are being planned and should start in late 2013. The results will be reported in a forthcoming book: <u>**Aging is Preventable**</u>.

The authors therefore conclude that we can become immortal. Many researchers believe that the first immortal is alive today.

Chapter 7
Are there Other Intelligent Lives in the Universe

"It is unnatural in a large field to have only one shaft of wheat, and in the infinite Universe only one living world."

Metrodorus of Chios
Greek Philosopher c. 350 BC

We know that all life on earth is connected in that they share significant percentages of DNA as previously described.

Initial life on Earth may well have been created or evolved as previously described. But many now think that life on Earth may have been "seeded" from a comet or other body from outer space.

In either event simple life is not a low probability event when the proper conditions are available.

We know that the universe was "created" about 13.7 billion years ago; Earth about 4 billion years ago.

There are billions and billions of planets that are very much older than Earth. They have had time for life to progress far beyond life on Earth.

Intelligent life on Earth began less than a million years ago. Life on older planets have had more than 4,000 times longer to develop and advance intelligent life than we have had here on Earth!

And if we on Earth got our seed from outer space, other planets may have gotten the same see. Life there could be akin to us. They would not likely look like us because all species on Earth developed from that initial seed and they look very different from each other.

Life on other planets would have developed based on their environments.

Let's visualize some possibilities.

Remember the sea worm we found near the volcano vents in the deep ocean?

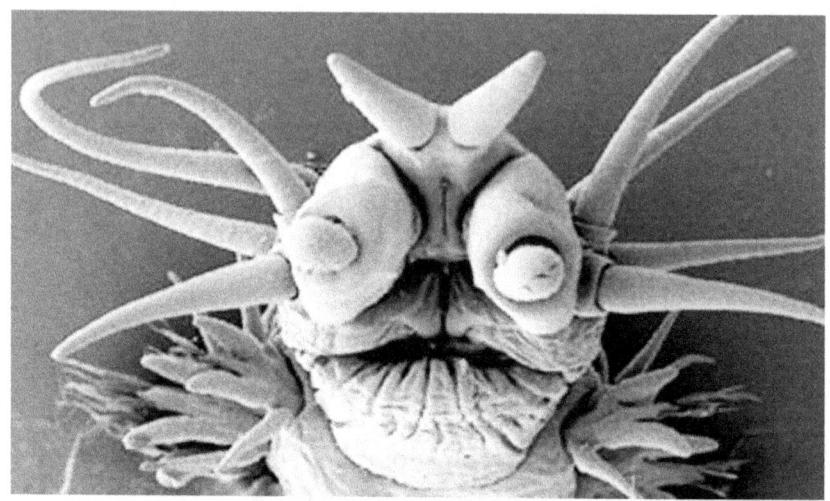

The shape of this creature here on Earth evolved because of the high pressure and high heat near the deep ocean vents where he lives.

If an alien evolved in similar condition on another planet he may develop these same "protective" characteristics. His skin could be very thick and tough to withstand the environment.

But he would have to have delicate hands with opposable thumbs to develop and use tools that a highly intelligent creature would have to have. He would also have to be bipedal to have his hands free for use. These are the hallmarks for development of high intelligence.

He could have almost any appearance but he would most likely have a body structure something like our body structure to have evolved to become an intelligent creature.

After all planet Earth "experimented" with the vast array of life forms that exist here on Earth and selected our form to evolved to be the most intelligent.

Intelligent life on other planets would most likely need to have body characteristics somewhat like ours to become intelligent.

Aliens less intelligent than us are not likely to be able to visit Earth. We will have to wait until we visit other planets before we encounter the beast-like ferocious creatures depicted in the movies.

So do we believe that there are other intelligent lives in the universe? Yes, we firmly so believe because:

1) There are so many billions of planets;

2) "Creation" of life is not an unlikely event when given the needed conditions;

3) We have seen that the environment on early Earth easily met the conditions and the continuing environments on Earth, while changing, have continuously been suitable for the creation of new life;

4) Billions of other planets have existed for periods up to 4,000 times the period in which we humans developed here on Earth. The statistics for intelligent live on other planets are just too high for it not to have occurred.

But how will we really know? Will we ever be able to prove it?

Chapter 8
Will we Ever Communicate with Otherworld Aliens

"If aliens ever visit us, I think the outcome would be much as when Christopher Columbus first landed in America, which didn't turn out very well for the Native Americans."
Stephen Hawking

The size and age of the universe lead us to believe that many technologically advanced civilizations must exist. But we have been searching the heavens for ages and have found no evidence.

While simple and even complex life is almost certainly probable throughout the universe; intelligent life capable of generating electromagnetic communication signals may be rarer.

Our communication signals became significant and began to be broadcasted in about 1949 and could have reached out a distance of 64 light years by now. Our Milky Way Galaxy is about 100,000 light years in diameter.

Thus we have broadcasted to less than 0.00004 percent of our galaxy and to nothing beyond our galaxy.

And even if some one, or something, detected our signals it would have taken up to another 64 years for us to receive a signal back from the ones further away that received our signals. Therefore the earliest we could expect a signal from them is 1949 + 64 = 2013.

Stand by; this may be the year in which we hear from intelligent alien life!

Mankind has been asking "are we alone" since before the beginning of civilization.

In the late twentieth century scientists agreed upon the basic idea of scanning the sky and "listening" for non-random patterns of electromagnetic emissions such as radio or television waves in order to detect another possible civilization somewhere else in the universe.

In late 1959 and early 1960, the modern SETI (Search for Extra Terrestrial Intelligence) era began when Frank Drake conducted the first such SETI search.

NASA joined in SETI efforts at a low-level in the late 1960s and early 1970s. Some of these SETI-related efforts included Project Orion, the Microwave Observing Project, the High Resolution Microwave Survey, and Toward Other Planetary Systems. On Columbus Day in 1992, NASA initiated a formal, more intensive, SETI program. Less than a year later, however, Congress canceled the program.

But a reduced version of SETI continues with funding from private sources and various companies. Their efforts rely also on "borrowing" time from various instruments that continue to search the sky.

The first SETI conference took place at Green Bank, West Virginia in November 1961. The ten attendees included Frank Drake and the popular astronomer and author Carl Sagan.

Frank Drake created the Drake Equation which considers an array of parameters to calculate how many intelligent civilizations may exist in our Milky Way Galaxy. Depending on the values of the various parameters he got a range of answers from just us to **182,000,000 intelligent civilizations** in just the Milky Way Galaxy.

Most everyone concluded that at least millions of intelligent civilizations exist in our galaxy alone and since there are billions of galaxies there are likely billions of intelligent civilizations.

But none of them beyond the .00004 percent of our galaxy will know about us, i.e. receive our communication signals until at least later this year.

The authors are patiently waiting.

But meanwhile, why have we not been able to receive their signals?

It may be that their signals are just not strong enough to reach us with significant signal to noise ratios for us to detect. There is a lot of "noise" coming from all directions in the universe. And the distances are literally astronomical.

This separation in time and distance may be a good thing.

The well known physicist Stephen Hawking wrote his famous book: A Brief History of Time. In the book he suggests that "alerting" extraterrestrial intelligences of our existence is foolhardy, citing mankind's history of treating his fellow man harshly in meetings of civilizations with a significant technology gap. He suggests, in view of this history, that we "lay low".

The concern over hostile alien life was raised by the science journal Nature in an editorial in October 2006, which commented: "It is not obvious that all extraterrestrial civilizations will be benign, or that contact with even a benign one would not have serious repercussions".

The result of extraterrestrial contact will be strongly governed by the benevolence or malevolence of an extraterrestrial civilization; by how advanced it is technologically; by whether or not such a species sends robotic probes to contact humanity, as opposed to radio signals from a centralized source; and by biological similarities and differences between humanity and the extraterrestrial species.

So do we really expect to soon have contact with intelligent alien life?

They could become aware of us, from our broadcast signals as early as this year but more likely within a few years.

Are they likely to contact us when they become aware of us?

Why would they?

Do we have anything that they want; anything worth traveling light years to get to us?

It most likely depends primarily on how many other intelligent civilizations they have already visited and how close we are to other civilizations.

The Authors tend to agree with Hawkins about the potential danger from aliens; but cannot help but desire contact.

If we continue our SETI activities we may eventually have the answers.

But most of us are too impatient to wait for eventually.

Chapter 9
Can We Travel Through Deep Space

Current spaceflight depends on rockets that burn fuel and an oxidizer. The exhaust velocity of these chemical based propellants cannot exceed much more than about 10,000 miles per hour. This technology is not suitable for space travel. It would take the better part of a year just to travel to our close neighbor Mars.

This technology is totally unsuitable for deep space travel.

Fortunately better technology is on the horizon. The plasma rocket ionizes fuel instead of burning it. The rockets use inert gases like xenon or krypton and an electrical source to accelerate the ions in the gas to create plasma. The higher the voltage exciting the plasma, the more velocity a rocket can achieve.

NASA has begun using a version of this kind of propulsion system for non-human space exploration, with solar arrays providing a limited but steady source of electricity for space missions that last years.

Ion Propulsion used for Deep Space 1 Probe

Most of the next travel planned will most likely be with unmanned space probes. Such probes have already visited every planet in our solar system except Pluto, and a mission there is currently being planned.

Future generations of ion engines, since there is no impediment to applying thousands of volts to charged particles, will be able to have an exhaust moving at several tens of thousands of miles per hour; theoretically much, much faster.

An Ion Propulsion System Could Take Us to Mars in 39 Days

Compact and efficient nuclear reactors on board could provide the electricity for the ion engines that would be able to propelling spacecraft fairly swiftly to nearby bodies.

We would, however still need the power of a chemical rocket to break the bonds of Earth's gravity. We would most likely use the chemically powered rocket to get to an Earth orbiting docking station and then switch to the ion engine for the distant trips to other planets.

Variable Specific Impulse Magneto-plasma Rocket (VASIMR)

Another candidate for space flight is the Variable Specific Impulse Magneto-plasma Rocket (VASIMR).

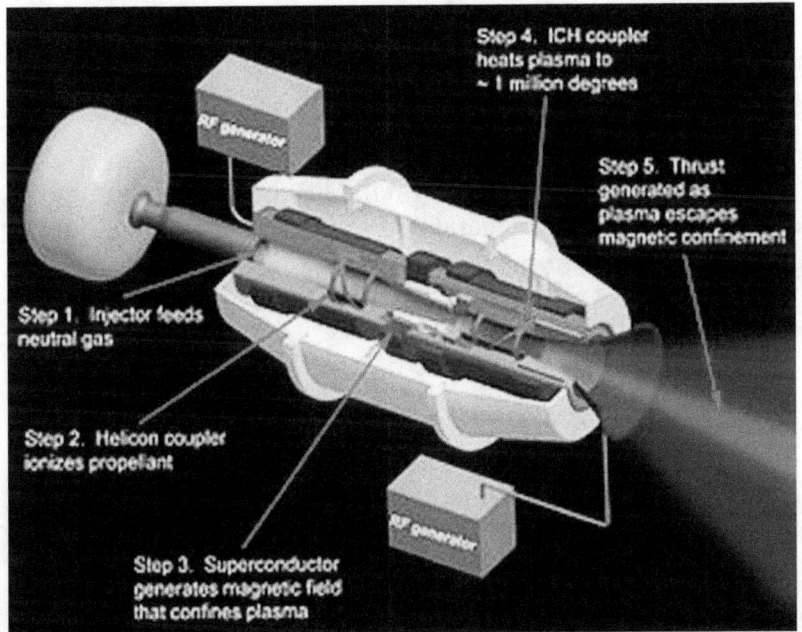

Step 4. ICH coupler heats plasma to ~ 1 million degrees

Step 5. Thrust generated as plasma escapes magnetic confinement

Step 1. Injector feeds neutral gas

Step 2. Helicon coupler ionizes propellant

Step 3. Superconductor generates magnetic field that confines plasma

Variable Specific Impulse Magneto-plasma Rocket (VASIMR)

Fuels for the VASIMR engine could include hydrogen, helium, and deuterium.

The use of hydrogen as the fuel for the VASIMR project has the side benefit of being able to refuel at its planet of destination. Hydrogen exists almost everywhere that we would likely want to travel.

NASA researchers therefore believe that VASIMR would not only allow for faster space travel, but once there it could refuel on the destination planet for the return flight to Earth.

In the less-distant future, VASIMR could even help keep the International Space Station (ISS) in orbit without requiring extra fuel to be brought up from Earth.

Another benefit of hydrogen fuel is that hydrogen is the best known radiation shield, so the fuel for the VASIMR engine could also be used to protect the crew from harmful effects of radiation exposure during the flight.

Electrical power sources for the VASIMR engine could include such things as a nuclear power system or solar panels. For long-range flights, the best option is nuclear power. Nuclear power is considered a must for travel to Mars.

VASIMR could be integrated with the Prometheus proposal to develop nuclear power generators for not to distant spaceflight.

Prometheus

After VASIMR completes some additional earthbound testing, its designers hope for it to be tested in orbit onboard the International Space Station.

Hydrogen is generated as a waste product on the current Space Station. This waste hydrogen could be used by the VASIMR engine to maintain the International Space Station's orbit without requiring any additional fuel.

Although VASIMR is still years away from being used in space, it has already shown great promise during tests on Earth.

It is entirely possible that the engine that will carry the first person to Mars is already running in a laboratory here on Earth.

Quantum Space Propulsion System

An even more advanced propulsion system for deep space travel is one invented by the young physicist Aisha Mustafa. Her system is based on the Casimir effect of quantum physics and could see mankind boldly go where no man has gone before.

The Casimir effect was first predicted over 40 years ago. An experimental verification of the dynamical Casimir effect was first achieved in May, 2011 in Gothenburg, Sweden.

Their scientists succeeded in creating light from the vacuum.

The scientists captured some of the photons that are constantly appearing and disappearing in the vacuum of space.

The experiment is based on one of the most counterintuitive, yet, one of the most important principles in quantum mechanics: that vacuum is by no means empty nothingness. In fact, **the vacuum is full of various particles that are continuously fluctuating in and out of existence.** They appear, exist for a brief moment and then disappear again. Since their existence is so fleeting, they are usually referred to as virtual particles.

It is believed that vacuum fluctuations may have a connection with "dark energy" which drives the accelerated expansion of the universe, as we discuss later.

The discovery of this acceleration was recognized in 2012 with the awarding of the Nobel Prize in Physics.

In quantum field theory, the Casimir effect and the Casimir–Polder force are physical forces arising from a quantized field. The typical example is of two uncharged metallic plates in a vacuum, placed a few micrometers apart as in a regular capacitor but without any external electromagnetic field.

119

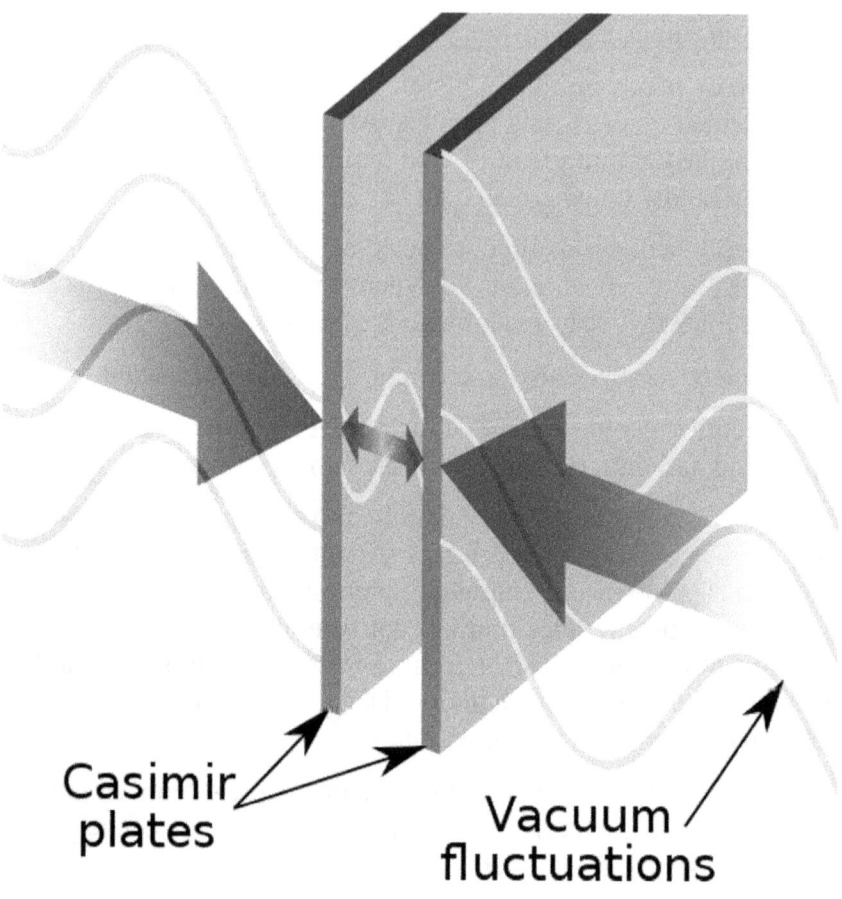

Casimir plates

Vacuum fluctuations

In order to fully understand how this works you have to have some knowledge of quantum physics, and even more demanding, you have to accept that energy and materials can almost spontaneously come into being from nothing.

What?

Yeah, follow my simple explanation.

The vacuum of space is not truly a vacuum. It is filled with particle waves of energy. It is from this "nothing" of particle waves that constantly appear and disappear that the big bang arose. That is, all of the matter and energy in today's universe derived from a single point in the big bang. We'll talk more about this when we discuss the universe in later chapters.

Now there are an infinite number of these particle waves in every part of the vacuum in the universe. If you place 2 parallel uncharged plates side by side but with a specific space between them you will create a space between the plates that also have a smaller infinite number of particle waves.

We hope it doesn't bother you that we are describing two different infinities; one is just smaller than the other, but both are infinite.

Now we have selected the spacing between the plates such that longer wave lengths of energy cannot fit in the small space. Therefore the "pressure" of the larger infinity outside the plates is greater than the pressure of the smaller infinity between the plates.

The Casimir effect therefore tends to force the two plates together. This force is the basis of this new propulsion system.

Our simple explanation may sound complex, but we have saved you from the mathematics of quantum physics and the Casimir equations. You're welcome.

Analysts say that this new propulsion system is expected to make space travels easier, cheaper and faster in future.

A very significant feature of this propulsion system is that it does not use a single bit of any kind of fuel.

Doesn't all of this remind you of Star Trek?

Star Trek might be just a TV show for many of us, but some believe that such high-tech space exploration may become a reality sooner than you might think.

OK, back to reality.

IIow do we pay for space travel?

In between interplanetary missions our spaceship will be able to act as an advanced space station for conducting a myriad of tests and perhaps a factory where production would benefit from zero gravity and/or the vacuum of space.

It will also likely find a job in mining asteroids.

The answer to the question "can we travel through Deep Space" is yes, eventually.

We may have to become immortal, as previously described; to be around when eventually comes.

The technology is already advanced enough that we could start a project.

What we need is leadership like President Kennedy when he visualized going to the moon; and implemented the vision.

Chapter 10
How did the Universe Begin

We now know that we are "children of the universe" in that every atom in our bodies was created out in the stars of the universe. Could that be a factor in our desire to travel into deep space and search for our extra terrestrial cousins?

Let's explore how the universe began and let's trace the particles and atoms from the first creations in the infant universe all the way to our bodies.

This will be a crazy trip; we hope you will go with us.

In the beginning there probably wasn't one.

We have been told that the universe began with the big bang however this theory introduces the concept of a discontinuity, or a singularity where the equations of physics cease to work. This has been a problem for most of us and even Einstein questioned it.

The authors do not accept this theory because we cannot accept a discontinuity and we want to know what was happening before the big bang.

We will explain what we believe was happening before the big bang in the next chapter.

In this chapter we will start moments after the big bang and review the evidence for how the creation of the universe continued from the big bang.

Stephen Hawking Comments

Let's start with comments from Stephen Hawking who is arguably Britain's most famous living scientist. He is also recognized as one of the leading theoretical physicists of recent decades. Since his twenties, he has suffered from Amyotrophic Lateral Sclerosis or ALS, a degenerative condition commonly known as Lou Gehrig's disease.

Hawking had previously appeared to accept the role of God in the creation of the universe. Writing in his bestseller **A Brief History Of Time** in 1988, he said: "If we discover a complete theory, it would be the ultimate triumph of human reason, for then we should know the mind of God."

In his new work, **The Grand Design**, co-written with American physicist Leonard Mlodinow, he argues that the Big Bang, rather than occurring following the intervention of a divine being, was inevitable due to the law of gravity; and that a creator was not necessary.

The Grand Design disagrees with Sir Isaac Newton's belief that the universe must have been designed by God as it could not have been created out of chaos.

Hawking counters with: "Because there is a law such as gravity, the universe can and will create itself from nothing…. spontaneous creation is the reason there is something rather than nothing, why the universe exists, why we exist."

He also says that M-Theory is a form of string theory that will eventually prove his beliefs: "M-Theory is the unified theory Einstein was hoping to find".

Hawking says the first blow to Newton's belief that the universe could not have arisen from chaos without God was the observation in 1992 of a planet orbiting a star other than our Sun. "That makes the coincidences of our planetary conditions: the single sun, the lucky combination of Earth-sun distance and solar mass, far less remarkable, and far less compelling as evidence that the Earth was carefully designed just to please us human beings".

Hawking's coauthor Mlodinow says that Creation is no longer a big mystery: "Ours and many other universes were created spontaneously from nothing and all the universes have different laws of nature and we happen to live in one that has laws that are friendly to our existence".

Big Bang Theory of Creation

The prevailing theory of creation is that a tremendous explosion started the expansion of the universe about 13.7 billion years ago. This explosion is known as the Big Bang. At the point of this event all of the matter and energy, and time itself, and space itself was contained in an infinitely small single point.

This prevailing theory does not say what existed prior to this event.

But we are going to take a stab at it.

The Big Bang was not a conventional explosion like a dynamite explosion where debris accelerates from the point of the explosion. The Big Bang was an explosion of energy, space and time.

We will try to explain as we proceed.

Edwin Hubble Theorizes the Big Bang

The origin of the Big Bang theory can be credited to Edwin Hubble. Hubble made the observation that the universe is continuously expanding. He discovered that a galaxy's velocity is proportional to its distance from us. Galaxies that are twice as far from us move twice as fast.

Another consequence of his theory is that the universe is expanding in every direction. This observation means that it has taken every galaxy the same amount of time to move from a common starting position to its current position.

Edwin Powell Hubble was an American astronomer and is generally regarded as one of the most important observational cosmologists of the 20th century.

Edwin Powell Hubble

The universe has been continuously expanding since the Big Bang and thus there has been more and more distance between clusters of galaxies.

In addition to Hubble's understanding of the velocity of galaxies emanating from a single point, there is further evidence for the Big Bang.

Big Bang Background Noise Discovered

In 1964, two astronomers, Arno Penzias and Robert Wilson were trying to detect microwaves from outer space when they inadvertently discovered a noise of extraterrestrial origin. The noise did not seem to emanate from one location but instead it came from all directions at once. It became obvious that what they heard was radiation from the farthest reaches of the universe which had been left over from the Big Bang. This discovery of the radioactive aftermath of the initial explosion lent much credence to the Big Bang theory.

COBE Satellite

Even more recently NASA launched the Cosmic Background Explorer (COBE) satellite to measure the diffuse infrared and microwave radiation from the early universe.

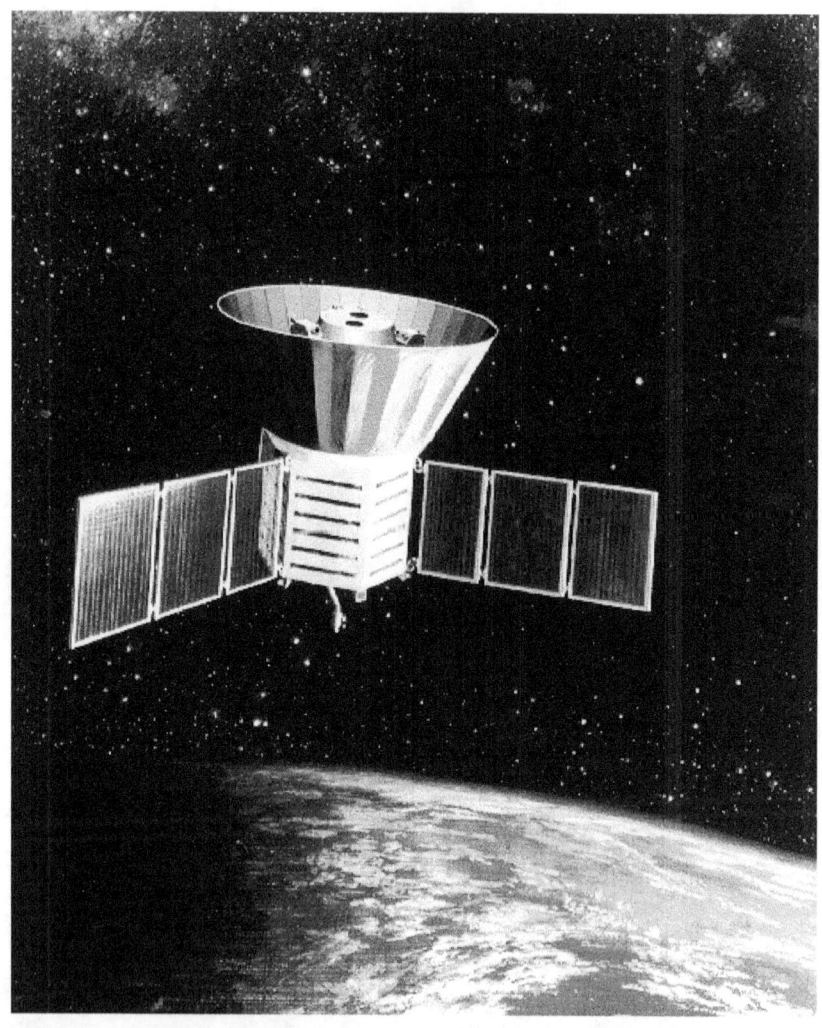

COBE was launched November 18, 1989 and carried three instruments, a Diffuse Infrared Background Experiment (DIRBE) to search for the cosmic infrared background radiation, a Differential Microwave Radiometer (DMR) to map the cosmic radiation sensitively, and a Far Infrared Absolute Spectrophotometer (FIRAS) to compare the spectrum of the cosmic microwave background radiation with a precise blackbody. Each COBE instrument yielded major cosmological discoveries.

Astro-2 Observatory

Then in June, 1995 NASA made additional astounding discoveries which lend themselves to the proof of the Big Bang theory.

Astronomers using the Astro-2 observatory were able to confirm one of the requirements for the foundation of the universe through the Big Bang. They were able to detect primordial helium, such as deuterium, in the far reaches of the universe. These findings are consistent with an important aspect of the Big Bang theory that a mixture of hydrogen and helium was created at the beginning of the universe.

We therefore have very good proof that the Big Bang started the expansion of the universe. But research will never be truly complete. Our hunger for knowledge will never be satiated.

What we continue to do and how much we learn is simply a matter of leadership.

Timeline for Expansion of the Universe

Let's follow a timeline for the expansion of the universe.

In the minuscule fractions of the first second after the Big Bang, what had been a complete vacuum began to expand into what we now know as the universe.

Immediately after the Big Bang a plasma soup formed.

The universe began to expand faster than the speed of light in a process known as Inflation, which we discuss in the next section.

The universe was tremendously hot as a result of particles of both matter and antimatter rushing apart in all directions. As it began to cool there existed an almost equal yet asymmetrical amount of matter and antimatter.

As these two materials are created together, they collide and destroy one another and become pure energy. Fortunately for us, there was an asymmetry in favor of matter. As a direct result of an excess of about one part per billion, the universe was able to mature in a way favorable for matter to persist.

As the universe expanded further, and thus cooled, common particles began to form, including photons, neutrinos, electrons and quarks which became the building blocks of matter and life as we know it.

At this time and because of the intense heat, there were no recognizable heavy particles such as protons or neutrons.

In about 4 hours after the Big Bang the universe had expanded to the current size of our solar system. The temperature was about 10 trillion degrees Kelvin.

After the universe had cooled to about 3 trillion degrees Kelvin, a radical transition began which has been likened to the phase transition of water turning to ice. Composite particles such as protons and neutrons became the atoms of matter during this transition.

About one to three minutes later the protons and neutrons began to react with each other to form deuterium, an isotope of hydrogen. Deuterium, or heavy hydrogen, soon collected another neutron to form tritium. Rapidly following this reaction was the addition of another proton which produced a helium nucleus.

Scientists believe that there was one helium nucleus for every ten protons within the first three minutes of the universe.

After further cooling, these excess protons were able to capture an electron to create common hydrogen. Consequently, the universe today is observed to contain one helium atom for every ten or eleven atoms of hydrogen.

By one billion years after the Big Bang dust and gas clouds began to form into stars. These first stars were very large and generally died out within 500,000 years of forming.

But stars continued to form.

By two billion years after the Big Bang galaxies began to form and by 3 billion years some galaxies merged to form larger galaxies. Such merging formed our Milky Way Galaxy.

Two Galaxies Merging

By four billion years after the Big Bang some of the super giant stars goes supernova and explode.

The debris from supernova includes the elements of which our bodies are made. **We are the children of supernova.**

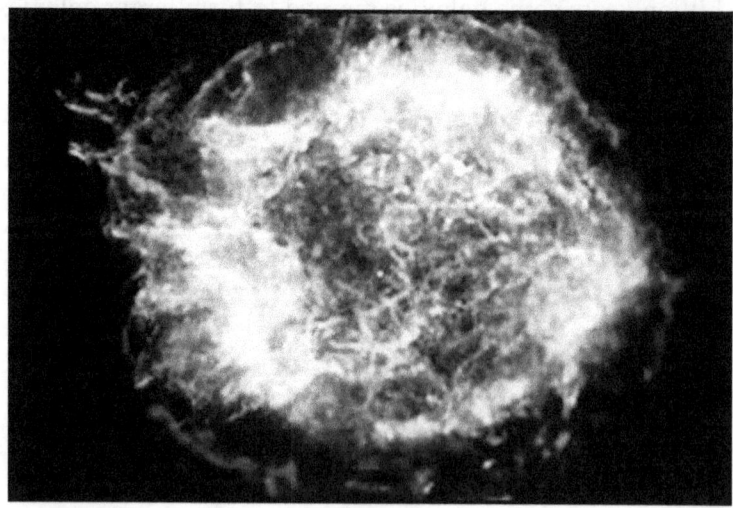

Supernova

Stars and planets continued to form.

Earth formed about 4 billion years ago, or about 9.7 billion years after the Big Bang. But the first stars and planets formed only about 1 billion years after the Big Ban.

We see therefore that some planets are at least 8.7 Billion years older than Earth. Such planets would potentially have civilizations much more advanced than us.

The Expanding Universe

The depiction below illustrates how the Big Bang started at a single point with no dimensions and rapidly expanded via inflation energy and then transition to "normal" expansion, and then began to transition to accelerated transition due to vacuum energy.

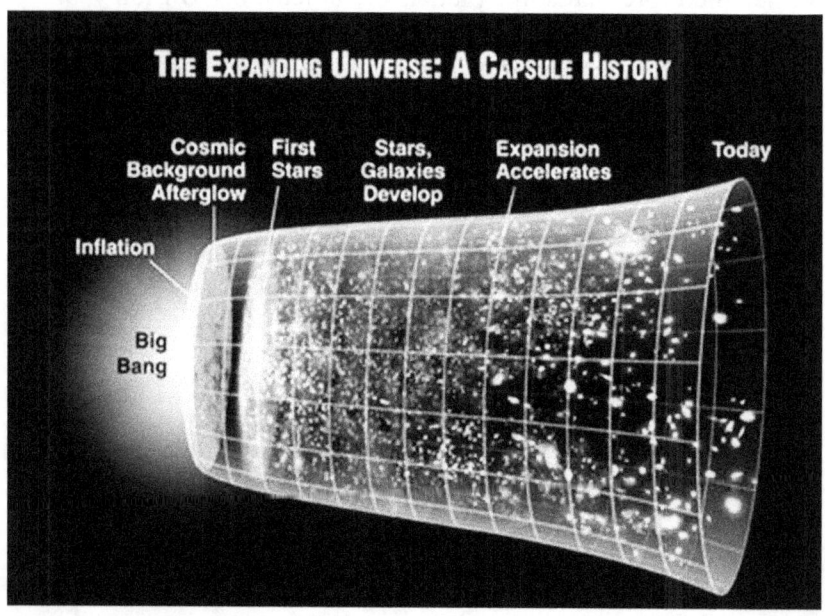

It is theorized that this accelerated expansion will continue with all galaxies and galaxy clusters becoming farther and farther apart until the universe approaches a complete vacuum.

Earlier we believed that the universe was static; neither expanding nor contracting. But Einstein's equations showed that the gravity of all the matter in the universe would exert a strong pull, pulling all the stars and galaxies toward each other and eventually causing the universe to collapse.

But then the Hubble measurements and theory, and other measurements proved that the Universe was not only continuing to expand, it was also accelerating the expansion!

There have been a multitude of theories and measurements to try to explain this. The authors believe that the answer is **vacuum energy**.

Vacuum Energy, Dark Energy

We mentioned this in the Chapter on propulsion systems for space travel. We need to repeat some of the explanation of vacuum energy here to better tie it to the creation of the universe.

Vacuum energy comes from the act of space itself producing energy; and this energy is "pushing" the universe outward.

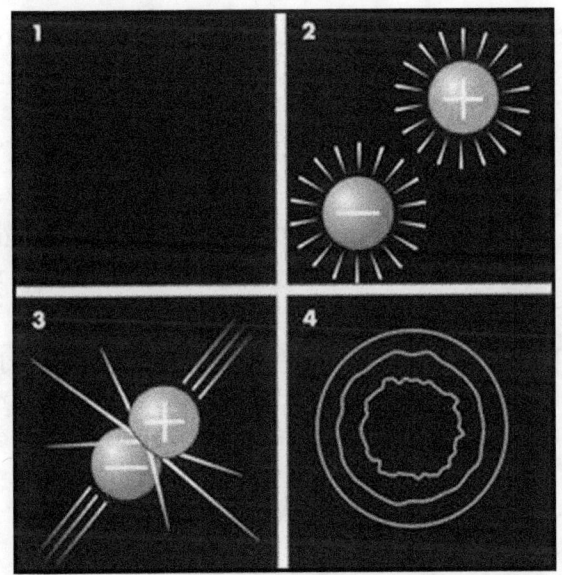

Vacuum energy can be illustrated with the chart where section 1 is empty space; section 2 shows two particles spontaneously appearing; section 3 shows the differently charged particles ramming together to annihilate each other; section 4 illustrates that they leave ripples of energy through space. [Diagram provided by Tim Jones on the Internet]

When dark energy from vacuum energy was discovered, it was realized that it could provide the "repulsive" force to explain the acceleration of the universe.

The process is that pairs of particles are constantly popping into existence throughout the universe. These "virtual pairs" consist of one particle with a negative charge and one with a positive charge. They exist for only a tiny fraction of a second before they collide and annihilate each other in a tiny burst of energy. This energy pushes outward on space itself causing the universe to accelerate apart faster.

One of the appealing elements of vacuum energy is that it could explain why the acceleration has only started fairly recently on the cosmic timescale.

In the early universe all the matter was packed much more densely than today such that there was less space between galaxies. So when everything was closer together, gravity was the dominant force and it slowed the acceleration of the universe that was imparted in the Big Bang.

Since there was less space in the universe, and the vacuum energy comes from space itself, vacuum energy played a much smaller role in the early universe.

Now, 13.7 billion years after the Big Bang the universe has grown much larger and the galaxies are not packed so close together. Their gravitational pull on each other is weakened. And there is much more vacuum energy because there is more space from which it can create, allowing the vacuum energy to play a more dominant role.

It has been calculated that this energy, now termed dark energy constitutes about 74 percent of all the matter and energy in the universe!

It should also be noted that dark energy shows itself only on the largest cosmic scale via its influence on individual galaxies and on the universe at large.

Dark Energy - Inflation Energy

It is believed that this Dark Energy is somewhat similar to the energy that caused the inflation immediately after the Big Bang. We discuss it in more detail in the next Chapter.

Visible Matter

Our universe is estimated to contain 100 billion galaxies, each with billions of stars, great clouds of gas and dust, and untold numbers of planets and moons and other bits of cosmic flotsam. The stars produce an abundance of energy from radio waves to X-rays which streak across the universe at the speed of light.

Yet all of this energy and matter that we can see accounts for only about four percent of the total mass and energy in the universe!

Dark Matter

Dark Matter makes up the remaining 21% or so of the universe. It is the most common matter in the universe yet it doesn't shine or reflect light. We can't even see it.

It is an invisible substance composed of particles that are far different from those that make up the universe's normal matter that we find in stars and galaxies.

Although astronomers cannot see dark matter, they can infer its existence by its gravitational effects in galaxies.

We have described how the Big Bang produced a soup of particles and energy and how some of these products became subatomic particles that later became atoms and the elements of all mater.

But not all of the products of the Big Bang became matter as we know matter. Some became WIMPs (Weakly Interacting Massive Particles). WIMPs are the subatomic particles which did not progress into ordinary matter. They are "weakly interacting" because they can pass through ordinary matter without any effects. They are "massive" in the sense of having mass (whether they are light or heavy depends on the particular particle).

WIMPs include neutrinos, axions, neutralinos and most likely many other particles which have not yet been found or even theorized.

Collectively WIMPs and such other particles have been termed Dark Matter and make up 21% of the matter and energy in the universe.

The collective mass of WIMPs tends to add to the attractive forces of matter and to oppose the repulsive force effects of Dark Energy.

The Dark Matter releases no detectable energy but exerts a gravitational pull on all the visible matter in the universe. So dark matter pulls matter inward and dark energy pushes it outward.

Astronomers first discovered dark matter while studying the outer regions of our Milky Way Galaxy.

Astronomers using NASA's Hubble Space Telescope got a first-hand view of how dark matter behaves during a titanic collision between two galaxy clusters. The wreck created a ripple of dark matter, which is somewhat similar to a ripple formed in a pond when a rock hits the water.

The ripple or ring's discovery is among the strongest evidence yet that dark matter exists.

The ring-like structure is evident in a composite image of the cluster made from Hubble observations. The ring can be seen in the blue map of the cluster's dark matter distribution, which is superimposed on an image of the cluster.

**Hubble Space Telescope Image of a
Distant Galaxy Cluster Surrounded by a
Ring of Dark Matter
[NASA/Ford (Johns Hopkins)]**

Another indication of Dark Matter is its effect on the rotation of stars within our Milky Way Galaxy. The Milky Way is shaped like a disk that is about 100,000 light-years across. The stars in this disk all orbit the center of the galaxy. The laws of gravity say that the stars that are closest to the center of the galaxy, which is also its center of mass, should move faster than those out on the galaxy's edge.

Yet when astronomers measured stars all across the galaxy, they found that they all orbit the center of the galaxy at about the same speed. This suggests that something outside the galaxy's disk is tugging at the stars. It is concluded that it is dark matter.

The same effect is seen in many other galaxies. And clusters of galaxies show exactly the same thing.

Composition of the Universe

So the composition of the universe consists of 3 types of energy/matter: normal matter, dark matter and dark energy.

The **normal matter** came about via a transition; some call it "decay", from about 4 percent of the plasma soup at the beginning of the Big Bang. This is the matter that we can see and which makes up our galaxies, stars, planets and ourselves.

The **dark matter** transitioned, or decayed from the soup to become WIMPs and other matter that did not continue the transition to become normal matter. We cannot see this matter but can detect it by observing its gravitational effects on the galaxies.

The **dark energy** is the energy from the Big Bang plasma soup that did not transition or decay.

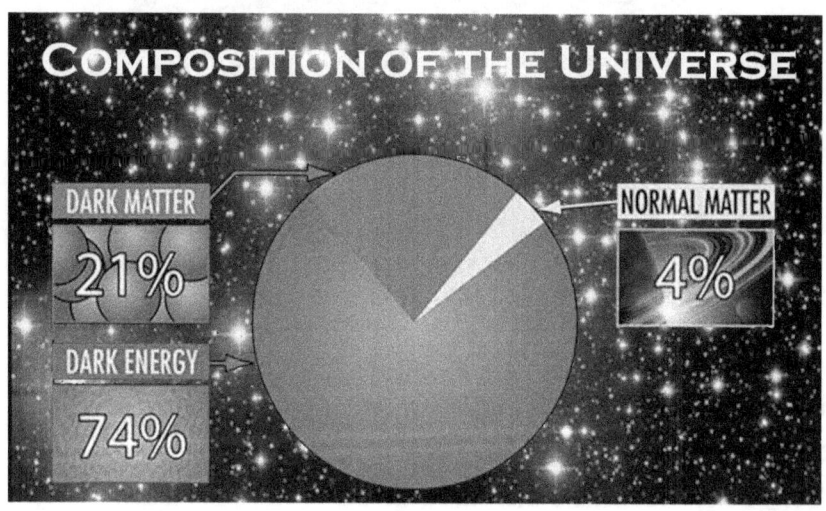

Size and Age of the Universe

The universe has expanded from an infinitesimally small point to its current size of about 93 billion light years in diameter.

The size and age of the universe can be fairly accurately approximated by applying the common physical equation of distance over velocity equaling time which uses Hubble's observations.

The two primary measurements needed are the distance of a galaxy moving away from us and that galaxy's red shift which indicates velocity.

Various observational analyses including comparative analysis of star illuminations and other factors were then used to derive at the distances to various stars and galaxies.

These measurement techniques have been used to calculate that the universe is 13.7 billions years old; i.e. it has been 13.7 billion years since the Big Bang.

Before the Big Bang

But what happened before the Big Bang?

Now that we know a little about vacuum energy and see how particles and energy can spontaneously appear and disappear in the vacuum of space, we can better speculate on what happened before the Big Bang.

Let's explore this in the next Chapter.

Chapter 11
Are there Parallel Universes

Yes, there are an infinite number of Parallel Universes.

To understand what they are and how they came to be we need to review some of the newer science that has been developed since most of us have graduated from high school and college.

Let's start with Einstein.

We know that Einstein's **Theory of Relativity** showed us that there are worlds beyond our 5 senses. Our 5 senses developed here on Earth for us to cope with and live in the simpler world as that described by Newton's laws of physics.

Einstein showed that **Newton's Laws** were just a subset of his broader theory. We also know that Einstein's theory did not fit nor take into account **Quantum Physics** which describes how the subatomic world works.

Einstein spent the last years of his life seeking the **Theory of Everything** which would unite the Theory of Relativity and Quantum Physics.

Then along came **String Theory** in 1969 but there were problems that were subsequently resolved by scientists coming up with five different superstring theories. These 5 distinct superstring theories then collectively evolved into Membrane Theory or **M-Theory**.

Stephen Hawking and Leonard Mlodinow, in their popular scientific book, **The Grand Design**, which we previously mentioned, take a philosophical position to support a view of the universe as a multiverse and suggest that M-Theory is the only candidate for a complete Theory of Everything.

According to Hawking in particular, "M-Theory is the only candidate for a complete theory of the universe."

The authors accept that M-Theory is the very best theory to explain the creation of our universe, and indeed the creation of an infinite number of parallel universes.

In the early years of the 20th century, the atom had long been believed to be the smallest building-block of matter. And this is what we were taught in school.

But it was later proven that the atom consisted of many smaller components called protons, neutrons and electrons.

Then beginning in the 1960s other subatomic particles were discovered. In the 1970s, it was discovered that protons and neutrons and various other particles are themselves made up of even smaller particles called Quarks and Gluons.

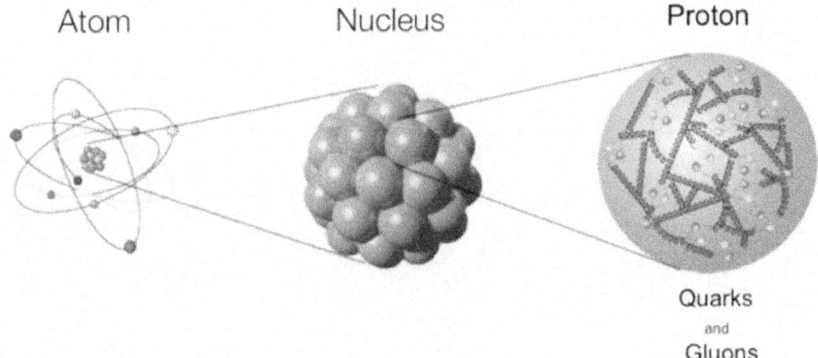

Atom   Nucleus   Proton

Quarks
and
Gluons

The Atom is depicted with a nucleus center of Protons and Neutrons Orbited by electrons with the Proton made of Quarks and Gluons.

Quantum theory was then devised to provide the set of rules that describes the interactions of these and other subatomic particles.

M-Theory

M-Theory goes deeper and shows that all particles and all the forms of energy in the universe are composed of one-dimensional "strings". These infinitesimal building blocks have only the one dimension of length; no height nor width.

M-Theory also states that the universe goes beyond the dimensions of our day to day world and has 11 dimensions.

Height, width, and length constitute three-dimensional space, and time gives a total of four **observable dimensions**. There are an additional 6 dimensions which we cannot detect directly.

Super-gravity theory then played a significant part in establishing the necessity of the 11th dimension. This 11th dimension was necessary to combine the differing theories of the 5 versions of string theories to arrive at M-Theory.

These "strings" vibrate in multiple dimensions, and depending on how they vibrate, they might exist in three-dimensional space as matter, light, or gravity or energy. It is the vibration of the string which determines whether it appears to be matter or energy (or gravity), and every form of matter or energy is the result of the vibration of strings.

We will try to describe the 11 dimensions of M-theory in simple terms.

We are all familiar with the first 3 which are the spatial dimensions of up-down, left-right and forwards-backwards. The fourth dimension is time.

We would therefore need 4 numbers to define a point where we or something may be located here on Earth in our Newtonian world: the coordinates of latitude, longitude and altitude plus the time.

M-theory physicists have concluded that there are 7 additional spatial dimensions. We therefore need 7 additional numbers to precisely define a point in the total world of M-Theory.

These additional 7 dimensions differ from our familiar spatial dimensions in that they are curled into small loops within the familiar spatial dimensions. This makes them periodic (i.e. if you walk in their direction you will after a while end up where you started), and they are very tiny. You can imagine them as small loops of additional space at every location in our regular 4 dimensional spacetime.

Each string has a length of only 10 to the minus 35 meters, many times smaller than the diameter of the nucleus of an atom. Any given subatomic particle is made of a string that vibrates and rotates at the speed of light. A particular particle gets its unique identity from the manner in which the string rotates and vibrates according to the dynamics of Einstein's theory of general relativity. The frequency of vibration corresponds to the mass of the particle.

These loops cannot be seen because they are so very tiny. We typically travel in these dimensions all the time, but we don't notice it because we are so very large compared to their tiny size.

Briane Greene used a good analogy of an ant on a garden hose. For the ant the garden hose seems 2 dimensional (3 if you count time); the ant can move forward and backwards on the hose, but it can also move around along the hose in a circular fashion. But if you stand some distance away the hose appears to be a line, a 1 dimensional object. You do not see the second curled-up dimension because it is too small because you are too far away.

Scientists are looking forward to "seeing" these small loop dimensions in the near future with some of the modern particle collider experiments such as with the LHC, since these can be used to probe the small distance scales where these additional dimensions exist.

OK, we know that M-Theory clashes with our intuitive geometrical notions. But stay with us.

The micro-atomic sizes of things and spaces at the tiny levels of M-Theory have no real geometric interpretation with our senses developed for the macro world in which we live.

Another analogy with objects in our daily world may be helpful in understanding M-Theory. Let's consider the surface of a lake. We can describe the waves that propagate on the surface of this lake and are able to say if we are above the surface or below the surface.

But when we look at the lake with very high resolution we start seeing the individual water molecules. Then the surface of the lake becomes a lot less sharp. In fact, the surface is not a well defined concept at the atomic level. There are molecules constantly leaving the water into the air and vice-versa.

We can only talk about the surface of the lake at macroscopic distances. The same happens with spacetime, we can use the standard geometric description only at long enough distances.

So M-Theory predicts that just as in the case of the surface of the lake, spacetime will loose its meaning at very short distances.

This does not mean that we cannot describe the system. In the case of the water, we have a well defined description of the system in terms of the water molecules. Similarly, in the case of spacetime we have a good description of what is going on but the necessary description uses less intuitive parameters and variables.

M-Theory relies upon extremely complicated math to arrive at its portrayal of reality. Such detail math is not appropriate for this book.

Heisenberg Uncertainty Principle

To better get a grasp on Quantum Physics and M-Theory we need to get familiar with the Heisenberg Uncertainty Principle.

When physicists started studying the quantum level in detail they noticed some very peculiar things about this tiny subatomic world. The most notable was that the particles that exist on this level take on different forms arbitrarily. When scientists observed photons, which are tiny packets of light, they saw them acting as both particles and waves. Even a single photon exhibited this shape-shifting. Some times it is a particle and sometimes a wave.

This has come to be known as the Heisenberg Uncertainty Principle. This tells us that we can never be fully certain of the nature of a quantum object and its attributes like velocity or even location.

This is supported by the Copenhagen interpretation of quantum mechanics suggested by the Danish physicist Niels Bohr. He concluded that none of the quantum particles exist in one state or the other, but in all of its possible states at once. The sum total of possible states of a quantum object is called its wave function. The state of an object existing in all of its possible states at once is called its superposition.

Young Hugh Everett agreed with much of what the highly respected Bohr had suggested about the quantum world. He agreed with the idea of superposition, as well as with the notion of wave functions. But he further suggested that the reason we see two or more different versions of a particle is because it resides in more than one universe at any given time.

As unsettling as it may sound, Everett's interpretation has implications way beyond the quantum level. It undermines our concept of time as linear. It says that **we are observing particles in a parallel universe.**

Some possible assurance that the interpretation is theoretically possible came in the late 1990s from a thought experiment which was an imagined experiment used to theoretically prove or disprove an idea. The experiment indicated that the particle could indeed, at least theoretically, exist in more than one universe simultaneously.

This thought experiment renewed interest in Everett's theory and spawned many writings about the possibility of parallel universes.

Michio Kaku: Mr. Parallel Universe

The idea of Parallel universes has been explored and expounded up on by Michio Kaku who is now known as Mr. Parallel Universe.

Michio Kaku is the Henry Semat professor of theoretical physics at the City University in New York and the man who, in the late 1960s, co-founded the field String Theory with a complex series of equations that described the behavior of sub-atomic particles into a coherent whole.

Kaku explains all the implications. "String theory predicts the universe is like a soap bubble that is expanding and dying," he says. Billions of years from now stars will blink out; the night sky will be dark and the oceans will freeze over.

But we may have an escape route because our soap bubble co-exists with other soap bubbles; every time a black hole forms it may be creating a baby universe. The matter being sucked in may be blown out the other side, creating a white hole in a twin universe, which will expand very rapidly, like our own Big Bang.

We will soon know more because the theoretical is rapidly becoming practical as billions of dollars are now being spent to test the validity of M-Theory.

Kaku expects physicists at Cern to soon find evidence of the vibrating strings with the Laser Inferometry Space Antenna (Lisa) which is waiting for re-funding.

"Lisa is three satellites connected by laser beams which stretch 3 million miles across space. This may pick up shock waves from the instant of creation and maybe even pick up the "umbilical cord" of our universe."

"String theory is the only game in town," he says, "and you can't afford not to play it to the end."

M-Theory demonstrates that parallel universes exist. According to the theory, our own universe is like a bubble that exists alongside similar parallel universes. M-Theory supposes that gravity can flow between these parallel universes and that they can come into contact with one another.

When these universes interact, a Big Bang like the one that created our universe occurs.

Lisa to Detect Gravity Waves

Lisa, the Laser Interferometer Space Antenna, was invented to measure the gravity waves in space that are at the heart of the creation process that we have been discussing.

Ripples in the fabric of spacetime regularly zip across the universe from titanic cosmic events such as the mergers of super massive black holes millions to billions of times the mass of the sun.

These gravitational waves can be detected by Lisa which was created by a joint mission between NASA and the European Space Agency (ESA).

LISA involved three identical spacecraft trailing Earth in an orbit around the sun. Each spacecraft would have targeted the other two with lasers, forming a triangle of light with sides five million kilometers long. Over the five-year mission, the laser beams would have helped detect subtle disturbances in the arrangement of the spacecraft caused by the passage of gravitational waves.

These gravity waves are predicted by Einstein's theory of gravity. Lisa was designed to prove their existence.

The original LISA project would have cost about $2 billion. Cost overruns in other NASA programs were one of the causes for cancellation of the project; however the teams are continuing to work to get the project funded. Many believe that the project will have to be reduced in scope, and costs to get funding.

As will all such projects, LISA will be a matter of political leadership.

The Future of Research

While grand projects are waiting for inspired political leadership, CERN and others are proceeding with basic research at the subatomic level.

CERN is the **European Organization for Nuclear Research**. It was founded in 1954 and is now one of the world's largest and most respected centers for scientific research. It conducts fundamental physics on subatomic particles (and waves).

It is most likely our best hope for continuing to learn what the Universe is made of and how it works.

The principal instruments used at CERN are particle accelerators and detectors. Accelerators boost beams of particles to high energies before they are made to collide with each other or with stationary targets. Detectors observe and record the results of these collisions.

The CERN Laboratory sits astride the Franco–Swiss border near Geneva. It was one of Europe's first joint ventures and now has 20 Member States.

Its Large Hadron Collider (LHC) is a gigantic scientific instrument located about 100 meters underground. (Subatomic particles are called Hadrons.)

The LHC is expected to revolutionize our understanding of the minuscule world deep within atoms and their effects on the vastness of the Universe.

For decades scientists have relied on the so called **Standard Model** which resulted from theories and discoveries of thousands of physicists over the past century. It was then basically agreed upon in the early seventies. This model provided a remarkable insight into the fundamental structure of matter: everything in the universe is found to be made from twelve basic building blocks called fundamental particles, governed by four fundamental forces.

This Model has successfully explained a host of experimental results and has precisely predicted a wide variety of phenomena.

But now we know that it does not tell the whole story. It did not and cannot include M-Theory.

Scientists believe that the new knowledge we will obtain from the high energies reached by the LHC will help us to build a new Model that will include M-Theory, and perhaps even further insights into how the Universe was Created and more details about the space-time continuum.

Parallel Universe Conclusions

So does all of this prove that parallel universes really do exist?

Not really. Our finite minds created for life here on Earth in our 3 dimensional world, 4 dimensions when we include time, just cannot visualize parallel universes.

The best we've been able to do is visualize an infinite number of "bubbles".

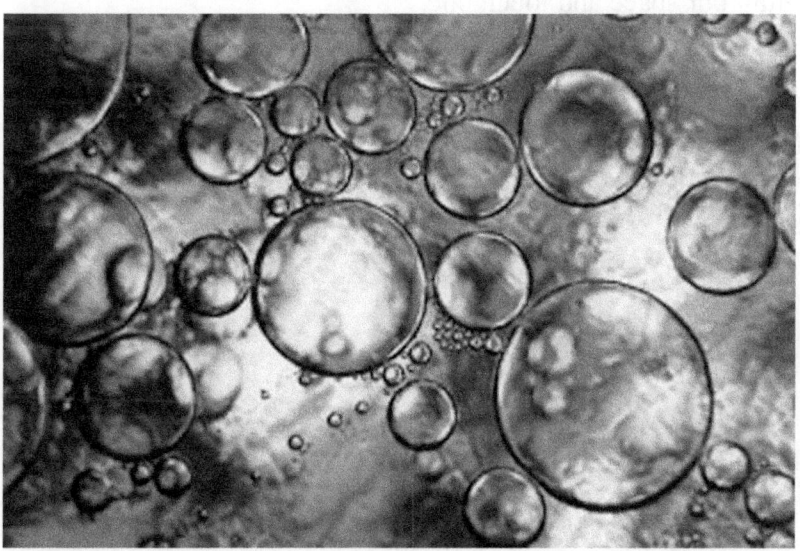

But theory says that they do exist so we accept the theory.

So where are they?

Well it is thought that these universes can be connected to our universe at times and unconnected at other times. Or you could say that some are connected and some are not at all times.

Some are within inches of our universe and some are much farther away than many multiples of the size of our whole universe.

Some of the universes abide by our laws of physics, some by other laws.

Some universes are very much larger than ours and some are microscopic.

There are an infinite number of parallel universes including all these situations and combinations thereof.

As we learn more about the subatomic world and how that world affects our universe we may eventually be able to travel to a parallel universe. We may also become able to just reach out within our space and touch one.

We have to recognize that much more is possible in the M-Theory world than the Newtonian world where we have been living.

If you want to take an imaginary trip into a Parallel Universe you may want to take a look at the fictional book: **Paranormal Portal to a Parallel Universe** previously mentioned.

Chapter 12
Is Time Travel Possible

Will we ever be able to communicate with or to travel to parallel universes? Will we ever be able to travel through time?

Let's explore some tidbits of information to see what we can conclude.

A Review of Relativity, Quantum Physics and M-Theory

Perhaps the greatest achievements of 20th century science is that all the laws of physics can be summarized by just two formalisms: (1) Einstein's theory of gravity, which gives us a cosmic description of the very large, i.e. galaxies, black holes and the Big Bang, and (2) the quantum theory, which gives us a microscopic description of the very small, i.e. the microcosm of sub-atomic particles and radiation.

But even the world's greatest physicists, including Einstein and Heisenberg, had failed to unify these two theories into one **Theory of Everything**. The two theories use different mathematics and different physical principles to describe the universe in their respective domains, the cosmic and the microscopic.

Fortunately we now have a candidate for this Theory of Everything: M-Theory.

M-Theory can explain the mysterious quantum laws of sub-atomic physics by postulating that sub-atomic particles are really just resonances or vibrations of a tiny string. The vibrations of a violin string correspond to musical notes; likewise the vibrations of a superstring correspond to the particles found in nature.

As a string moves in time, it warps the fabric of space around it, producing black holes, wormholes, and other exotic solutions of Einstein's equations. Thus M-Theory unites both Einstein's theory and quantum physics into one coherent, compelling picture.

The equations, although difficult, are well-defined.

And the equations coupled with observable evidence essentially prove that we live in an M-Theory world.

Wave Function of the Universe

Recall that in Quantum Physics a particle can be in more than one place at the same time. The sum of all the places that it can be is its wave function. We can detect it as a particle at any instance of observance. It may be in any location within its waveform.

The wave-function of a particle is essentially the function that determines its most likely location at any given time. Wave-functions are largest where that particle is observed to be, but also extend throughout the known universe in accordance with the sum-over-paths method.

Everything has a wave-function - elementary particles possess wave-functions and make up all other matter in the universe, so it is a logical conclusion that universes have wave-functions.

Large or small, exotic or common, all objects have wave-functions. Consider the wave-function of a ball sitting on a flat surface. Its location probability is largest where it is observed, i.e. where it sits. But its location also extends everywhere around us and in all places in the universe, and even in parallel universes. However, the likelihood of the ball suddenly appearing in any of these locations is infinitesimal. The likelihood of such changes in location depends on Planck's constant.

Perhaps we should remind you of just what Planck's constant is; it is the fundamental physical constant characteristic of the mathematical formulations of quantum physics which describes the behavior of particles and waves on the atomic scale.

The significance of Planck's constant is that radiation and other energy packets are emitted, transmitted, and absorbed in discrete energy packets, or quanta, determined by the frequency of the radiation and the value of Planck's constant.

The energy of each quantum equals Planck's constant times the radiation frequency. The dimension of Planck's constant is the product of energy multiplied by time which gives a quantity called action. Planck's constant is often therefore defined as the elementary quantum of action.

Hawking and his colleague Hartle propose to calculate the wave-function of the universe using the sum-over-paths method, which begins with the assumption that the universe has all possible histories. Moreover, they would calculate this sum in imaginary time, not ordinary time. This is because imaginary time travels at right angles to ordinary time and "meets" with the three spatial dimensions to create a smooth surface similar to the surface of the earth. This eliminates the singularities (points of infinite curvature) present in ordinary time, allowing the history of the universe to be reliably calculated. Also unlike ordinary time, imaginary time has no beginning or end, so progression through it is determined entirely by physical laws.

For Hawking and Hartle's calculation, you must begin with a wave-function describing all possible universes - an infinite number. The wave-function is large near our own universe and infinitesimal near others in which life is impossible or the known laws of physics do not apply.

Because of the wave-function's concentration in our own universe, it is the most likely of them all, but there is a chance - albeit vanishingly small - that an object from this universe would suddenly make a quantum leap into another one. Proving this conjecture mathematically is one of the primary goals of quantum cosmology, which applies quantum theory to the large structures of cosmology.

The Hawking/Hartle theory also postulates the existence of wormholes connecting the different universes as illustrated in the image below.

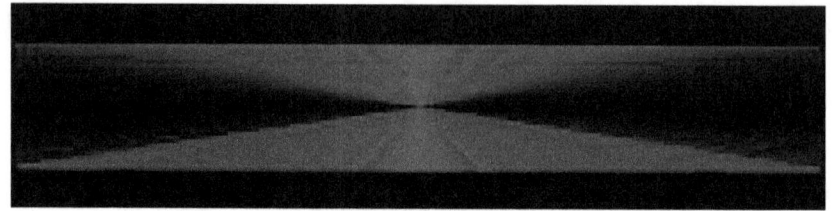

According to them, the multitude of universes should be connected by wormholes, as in the second image below. You can see that some of these universes are connected with many others, while others are isolated.

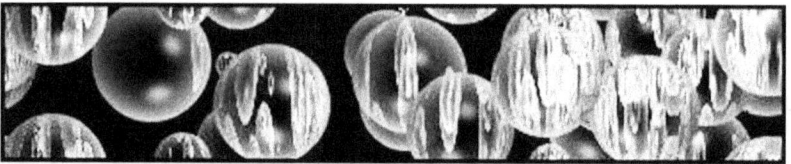

Stephen Hawking has speculated that the universe, indeed all universes can be like the wave function of a particle.

The mathematics of Hawking and Hartle's idea allows for communication between parallel universes via wormholes. Thus, their idea is both testable and directly relevant to our universe.

Wormholes

Wormholes connect two otherwise unrelated regions. They present paths much quicker to travel than the paths presented by ordinary space.

A beam of light traversing a path between two points in curved space-time can take longer to complete the journey than a hypothetical spaceship taking advantage of a wormhole's shortcut connection between the two distinct regions of space-time.

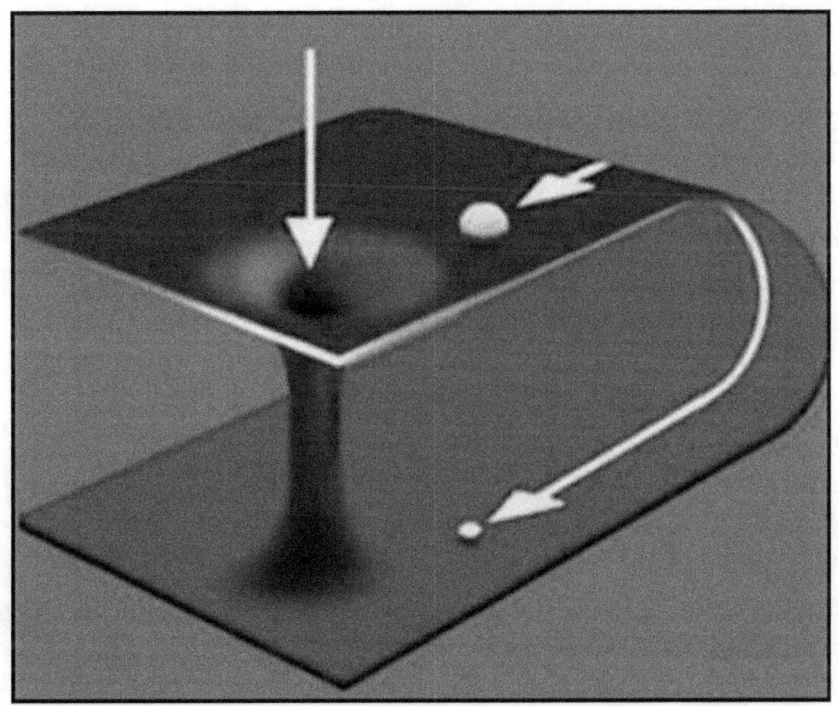

In 1935 Albert Einstein and Nathan Rosen realized that general relativity allows the existence of "bridges," originally called Einstein-Rosen bridges but now known as wormholes. These space-time tubes act as shortcuts connecting distant regions of space-time. By journeying through a wormhole, you could travel between the two regions faster than a beam of light would be able to if it moved through normal space-time.

As with any mode of faster-than-light travel, wormholes offer the possibility of time travel.

Wormholes may be possible on microscopic distances as a result of the quantum foam which allows particle/antiparticle pairs to leap into existence momentarily, then subsequently disappear.

Such microscopic distances may be at the points where parallel universes touch each other. Wormholes are therefore commonly postulated to exist where parallel universes touch. They are also postulated to exist at the centers of black holes as previously illustrated above.

Wormholes allow travel between different times as well as different locations. The associated mathematics makes time travel theoretically possible.

The relativistic description of black holes requires wormholes at their centers. These wormholes, called Einstein-Rosen bridges after Einstein and his collaborator Nathan Rosen, seem to connect the center of a black hole with a mirror universe on the "other side" of spacetime.

At first this Einstein-Rosen Bridge was considered just a mathematical oddity. But in 1963 Roy Kerr devised the famous Kerr solution to Einstein's equations which gave a realistic description of black holes.

Kerr assumed the star that would form the black hole to be rotating and found that it would not eventually collapse to a point, but rather to a ring. **The ring, rather than a point eliminates the need for the singularity and therefore travel through it could be theoretically possible.**

Sidney Coleman, a famous physics professor at Harvard, has recently put forward the theory that wormholes eliminate excess contributions to the cosmological constant, a measure of the inherent energy of vacuum. When Coleman added the contributions of the infinite series of wormholes proposed by Hawking, he found that the universe's wave-function grows very large around a value of zero cosmological constant, rendering the possibility extremely likely at this value.

More studies in this area could therefore lead to time travel.

Time and M-Theory

Another tidbit of information for Time Travel is provided by M-Theory which provides a possible means of traveling between parallel universes, at least for some subatomic particles. The matter in our universe is "attached to our membrane" and therefore would theoretically have no means of un-attaching and traveling to another universe. However in studies of neutrinos it was hypothesized that a special kind of neutrino called a sterile neutrino exists and it is not attached to our membrane. It would therefore be capable of traveling through the extra dimensions of parallel universes.

The sterile neutrinos could travel faster than light by taking shortcuts through the extra dimensions. Einstein's general theory of relativity allows, under certain conditions, for such faster than the speed of light travel.

Researchers posted a paper on this subject titled "**Neutrino Time Travel**" in 2007. (Hollywood grabbed onto the idea and made 2 movies based on the concept.)

Black Holes

Black Holes per se may also provide another tidbit of information for time travel.

A black hole is a massive dead star whose gravity is so intense than even light cannot escape. It continues to suck in matter to grow to an even more massive size.

NASA scientists have been studying a super massive black hole at the center of galaxy M87 which is 50 million light years from earth. Its colossal mass weighs more than 2 to 3 billion suns!

The best description of a spinning black hole was given in 1963 by the New Zealand mathematician Roy Kerr as mentioned above. Using Einstein's equations of gravity, Kerr speculated that matter sucked into the black hole could be shot out as a "white hole" in a parallel universe!

Hundreds of mathematical solutions to wormholes have now been found to Einstein's equations.

As these activities continue and eventually enable us to combine the tidbits of information from each, it appears that we will be able to define just how time travel could become possible.

Paranormal Events

Another tidbit of information for our consideration is the thousands of reported paranormal events.

Paranormal investigators have claimed to have found portals to a parallel universe. They say that beings from a parallel universe have visited our universe here on Earth. They have recorded orbs of light that they claim are such visitors.

In our research for this book one of us attended such a paranormal event where a paranormal investigation team said they would demonstrate that they could communicate with the dead that existed in a parallel universe.

I was more than a bit leery, but totally curious. I had to go.

Several of us went to a very weird old house where the Paranormal Investigative Team was working. They had some pretty sophisticated equipment.

One machine measured temperature and its rate of change. They explained that when "entities" came over into our world that they need energy to materialize. That's why it gets cold when ghosts, or whatever, come on the scene. They suck up the energy from the room or area so they can come into being.

The more energy that they are able to absorb, the more visible they become.

The team leader asked if we believed that people had souls. I told her that I was taught to believe that in my southern Baptist upbringing, but I had drifted away as I studied and worked in the physics based aerospace industry.

The paranormal study leader agreed that the idea of a soul certainly goes against the simple laws of physics that we live with day by day, here on earth. But she noted that it does not go against Einstein's Theory of Relativity.

She noted that Einstein showed us that there are many things beyond our five senses. Matter - the flesh of our bodies - and energy, like the energy we use to move about and live.... can be intertwined and transformed with and by time... time does not flow in the simple manner that we live with everyday. It can flow at very different rates.

That's relativity. On the other hand, Quantum Physics has shown us that the very atoms and sub-atomic particles of which our bodies are made can be in more than one place at the same time as we have previously described.

The notion of parallel universes is now widely accepted based on the mathematics of M-Theory.

Most scientists in the field believe that there are definitely worlds - universes - beyond the perception of our five senses. Calling them paranormal is as good as any other word.

The team proceeded to prove the reality of what they had been detecting. Their equipment recorded the event in some detail.

The temperature sensing equipment as previously mentioned showed significant and rapid temperature drops when the "entities" appeared.

Their electromagnetic sensors recorded the spikes of appearance and higher than normal levels as their presence continued.

Their infra red cameras detected image hot spots that varied in size from small orbs to life size wisps of bright energy.

Their acoustic equipment recorded static "noise" and varying levels of frequency specific hums with their presence.

Thousands of people have reported such paranormal encounters, and of seeing ghosts down through the ages.

I spent some time doing more research and then tried to speculate as to what could cause such paranormal events and ghost stories down through the ages.

I tried to find the science behind it. I described some of my speculations in the previously mentioned books: **Paranormal Portal to a Parallel Universe** and **Finding the Soul, Surviving Death**.

Time Travel Conclusions

So what can we conclude about time travel?

We already know that M-Theory provides us some insight into the strange world of subatomic physics where particles have wave functions that allow them to be in more than one place at any instant of time.

We also believe that the multiplex of parallel universes most probably have similar wave functions.

The mathematics of M-Theory gives explanations to all sets of observed data within both Einstein's Relativity and within Quantum Physics.

These same mathematics allow for faster than light travel; and time travel.

Will we be able to travel in time?

Our astronauts already have. When they returned from space travel they came back to a world older than they were. OK, it was just by fractions of seconds, but it was a beginning.

We fully expect to send particles through time travel relatively soon. And if Kerr's ring description of the black hole proves correct and Hawking's description of portals between parallel universes prove correct, then someday we will travel in time.

Epilog - Conclusions

OK, we have described what we believe about **Life and the Universe**. We addressed most of the key questions that have been asked down through the ages. We provided a lot of information; so much so that perhaps a summary is required.

Let's summarize what we have learned and what we believe.

Life

We believe that life is DNA because it is DNA that survives and perpetuates each species.

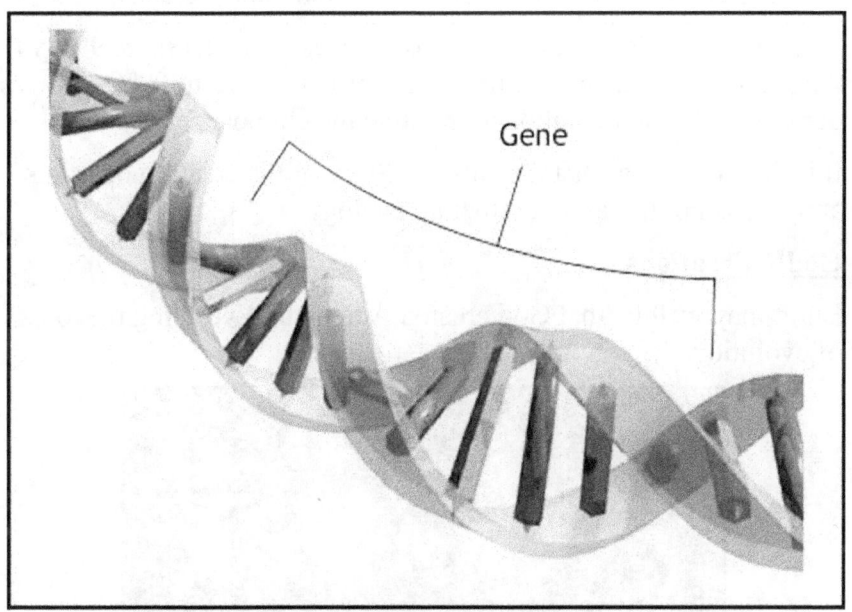

A Gene in the DNA Helix

The simple components of life may have come to Earth from comets and asteroids. It is however more probable that life here on Earth began from the spontaneous creation of amino acids as demonstrated in the Miller and Query experiment.

Continuous creation/evolution followed until the species of man was created.

This process began about **4 billion** years ago and continued with creation of various species until about **1.5 billion** years ago when the chain of creation split into the plant and animal kingdoms.

The creations in the animal kingdom continued until about **25 million** years ago when the line of the common ancestor of monkeys and apes split to form the great ape branch of creation.

Creation continued until a branch of the great apes produced the Hominoid branch about **6 million** years ago. This Hominoid branch continued to produce a series of pre-humans and then the first true Homo sapiens were born about **200,000** years ago.

As a result of this creation/evolution process humans share 99% of their DNA with the pre-human Neanderthal and 95 to 99% with our closest living evolutionary relative the chimpanzee.

It is therefore clear that the Biblical Story of the creation of Adam and Eve is not literal, but rather an analogy.

God's Creations

But it may well be that God created Adam and Eve using the tools of evolution.

So the question of whether there is a God that created us via evolution or whether we are mere happenstances of Darwinian time cannot be definitively answered. There is no clear evidence either way.

Man may well have a soul, or spirit that survives death. The more we learn about the relationship of matter and energy at the subatomic level, the easer it is to believe that our souls of pure energy could exist and survive death.

Will our souls eventually be with God?

Some people believe that God was created by Man to answer the unanswerable. The Authors believe that there is a God and that the concept of God will continue forever, if for no other reason, because Man will never have all the answers.

Immortality

If we define immortality as the prevention of aging, we may well become immortal even without a soul. The decoding of the human genome and other recent discoveries strongly suggest that we will eventually become immortal here on Earth. Key researchers believe that the first Immortals are living today.

Details will soon be published in the book: **Aging is Preventable**.

Other Intelligent Life

The Authors believe that the Universe is heavily populated with Intelligent Life because of the enormous number of planets and because we see that life can develop fairly easily in a diversity of environments.

Frank Drake created the Drake Equation which considers an array of parameters to calculate how many intelligent civilizations may exist in our Milky Way Galaxy. Depending on the values of the various parameters he got a range of answers from just us to **182,000,000 intelligent civilizations** in just the Milky Way Galaxy.

Most experts in the field conclude that at least millions of intelligent civilizations exist in our galaxy alone and since there are billions of galaxies there are likely billions of intelligent civilizations.

Why have there been no communication with us?

Remember the enormous size of the Universe. Essentially none of the intelligent civilizations even know abut us. Our puny electromagnetic transmissions of radio, television, etc cannot have yet reached more than .00004 percent of our galaxy and no civilizations beyond our galaxy.

We have been listening for communications and any form of electromagnetic transmissions from intelligent civilizations without any success. Now our primary listening programs have been cancelled buy congress due to lack of funds.

The Authors believe that it is just a matter of time before we hear some signals from or between intelligent civilizations, even with our reduced listening activities.

Space Travel

Maybe when we again get visionary leaders like President Kennedy who visualized and then implemented the trip to the moon, we will again become players in this field.

We have the technology to initiate space travel programs. We just need leadership.

The Universe

Our rate of learning about the Universe is truly explosive.

Largest Object Found in the Universe

Research, especially in particle physics from Europe, is providing the information needed to better understand how the Universe began and where it is going.

We have recently discovered:

- The matter, termed Normal Matter, that makes up our planets, stars, galaxies, out bodies and every thing that we can see makes up only 4% of the matter and energy in the Universe;

- Vacuum Energy which shows how particles spontaneously come into existence in a vacuum;

- Dark Energy that makes up 74% of the Universe;

- Dark Matter that makes up 21% of the Universe;

- Information that lets us speculate about how the Big Bang began and what came before the Big Bang;

- Information that can eliminate the need for the singularity or discontinuity in our laws of physics;

- M-Theory which combines Einstein's Theory of Relativy and Quantum Physics into the Theory of Everything.

Parallel Universes

Most researchers in the field now believe that parallel universes do exist.

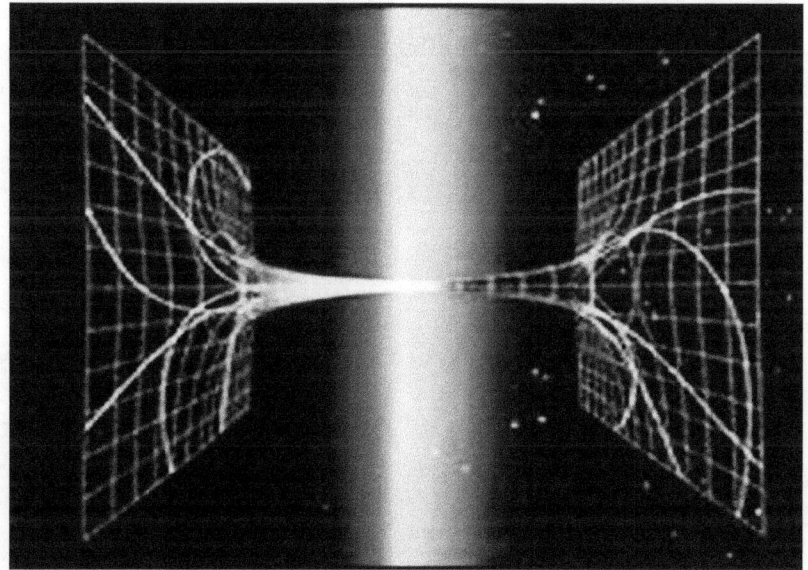

Illustration of Parallel Universes

M-Theory has gone much deeper than Einstein's Theory of Relativy and the principles of Quantum Physics show that all particles and all the forms of energy in the universe are composed of one-dimensional "strings". These infinitesimal building blocks have only the one dimension of length; no height nor width.

These "strings" vibrate in multiple dimensions, and depending on how they vibrate, they exist in three-dimensional space as matter, light, or gravity or energy. It is the vibration of the string which determines whether it is matter or energy (or gravity), and every form of matter or energy is the result of the vibration of these strings.

M-Theory also states that the universe goes beyond the dimensions of our day to day world and has 11 dimensions as we have explained in this book.

There are an infinite number of parallel universes. Our Universe is just one in this sea of infinity.

Current NASA projects will soon provide the proofs of some of the elements of M-Theory and help establish solid foundations for understanding more about parallel universes.

The Future – Where We Are Heading

Our accumulation of knowledge during the last 100 years has been astonishing. And the last few decades have been almost exponential. Our rate of learning will likely become even greater during the near future.

Consider how well off we are compared to all previous civilizations.

Try to visualize the world 150 years ago:

- There were no cars.

- There were no airplanes.

- There was no electricity.

- There was no television.

- There were no personal computers or Internet.

- Medicine was still fairly primitive with the average life span being 39 years (**average**, but many lived into their sixties).

- We knew essentially nothing about the Universe.

It was a very different world from today.

We are so much better off compared to those of yesteryear. We enjoy the fruits of the major advances in knowledge and technology that have occurred in the last 150 years. These advancements are expected to continue at an even faster rate in coming years.

Now try to visualize our world 150 years into the future.

- We may routinely travel to great Earth satellites and to planets and moons within our solar system.

- We may communicate with other intelligent civilizations.

- We will routinely eliminate aging.

- We will replace damaged body organs with robotic parts.

- Future immortal Earthlings with performance enhanced organ replacements may advance and dominate any "normal" humans.

- We may communicate, and possibly visit parallel universes.

This may be a very conservative view of our future. Can you believe that such radical changes may occur? Think again back to the world of 150 years ago as compared with today.

Eternity

When did time and every thing all begin and when will it end?

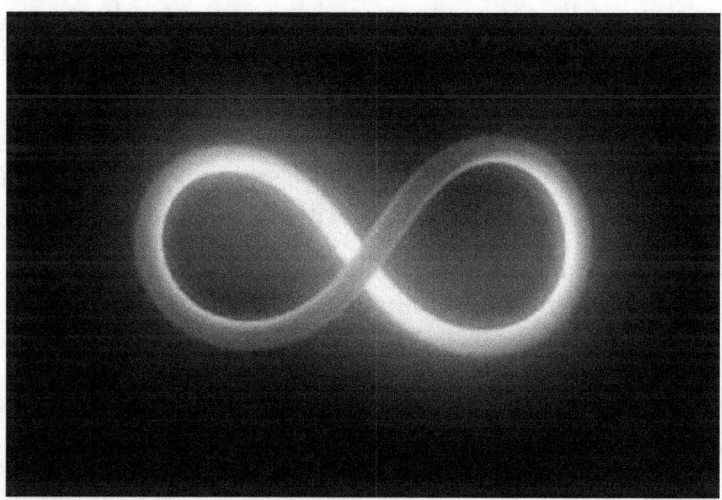

Our universe is about 13.7 billions years old as measured from the Big Bang. But that was not the beginning of time.

Our universe was created as one in an infinity of universes. And the process is most likely continuing.

The way we humans think almost requires a beginning and an ending for everything. We recognized the Big Bang as the beginning of our universe. However, we had to accept a singularity in our laws of physics to have this beginning.

M-Theory tells us that we do not need such a singularity, or discontinuity. Our universe is just one in a sea of infinite universes.

We have no "model" for the beginning of parallel universes. It is likely that there was no beginning and that parallel universes have always spontaneously popped into being via vacuum energy as we previously described.

We see no reason why this process would end.

It appears that the creation of parallel universes will continue for all eternity.

Our universe will continue to expand at a continuously increasing rate because more and more vacuum energy will come into being as there is more and more space, or vacuum in our universe.

As our universe continues through eternity it will approach a complete vacuum where all the planets, stars and galaxies are torn apart and dispersed into a vacuum with essentially no density.

Somewhere during this process our universe is likely to be absorbed or destroyed by another parallel universe.

That will be a very long time from now.

But we will have a problem long before that because our planet will cease to exist long before any universe-wide apocalypse. Our Sun is expected to expand into a red giant star and engulf Earth in about 5.4 billion years.

Before that time we will likely be able to move to another parallel universe and continue the moving process throughout eternity.

Mankind, or the more advanced species that our progeny will likely become will find a way to survive, perhaps for all of eternity.

Is There a God

All questions generally lead back to the fundamental question: Is there a God.

If Man created God to answer the unanswerable as some have believed, then as our knowledge increases the probability of a real God should diminish.

But as our knowledge increases we are becoming more and more compelled to believe that God must have created the very exotic and wonderfully varied universe and parallel universes that we are discovering.

As we learn more about our creation and the introduction of **informational management properties** in our design, and the manner in which information flows through and between cells and sub-cellular structures, we are compelled to believe that God installed these control mechanisms that give our cells the **functionality** needed for life.

As we learn more about how our brains have worked to create our minds we are compelled to believe that we had a creator.

Einstein's Theory of Relativity was totally incompatible with what we had understood and experienced in our daily lives here on Earth. His theory was considered far-out fantasy by most people. Yet in just a few decades it led to the knowledge to allow us to invent the atom bomb and atomic energy. Now most everyone believes his strange theory to be true.

Our discovery of Quantum Physics introduced the unbelievable idea that particles of which we are made could be in more than one place at any point in time. Fantasy again? Now we know it to be true and knowledge of this subatomic world has led to the explosion in electronic devices that have become major parts of our daily lives.

M-Theory led us to understand the spontaneous creation of vacuum energy in the vacuum of space. Now we do not have to rely on the notion of a singularity to understand how our universe was created or how our universe is just one in a sea of an infinite number of parallel universes.

The concept of God has also been viewed as a fantasy by many; even though most believe in God based on faith as taught by their religions.

But when will we have proof for the existence of God?

We thought we had achieved a great breakthrough in knowledge when we first "realized" that everything, including our bodies was made of atoms.

Then we experienced another breakthrough when we began to understand that the atoms were made of protons, neutrons and electrons.

Then another breakthrough when we realized protons and neutrons were made of quarks and gluons.

And now another breakthrough with M-Theory when we learned that every thing is made of vibrating strings.

Will we soon learn that the vibrating strings are just pure energy?

The Authors believe that God exists and that He is pure energy.

Pure energy came into being spontaneously via the process of vacuum energy as previously described. Some of this energy transformed into matter and thereby **created the universe and life**.

This visible matter makes up 4% of the universe.

Some of the energy did not make the full transition and became dark matter which makes up 21% of the universe.

Some of the energy did not make any transition and remained dark energy which makes up 74% of the universe.

We know that 4% of this energy in the universe transformed to **become** the universe and everything in it that can be seen. This energy became, i.e. **created the universe and us**.

The Bible calls this creating energy that made the universe God.

In the beginning God created the heavens and the earth.

Genesis 1

What should we call God; how should we describe Him?

*****

"If God did not exist, it would be necessary to invent him."

Voltaire
(1694 – 1778)

"Science without religion is lame; religion without science is blind."

Albert Einstein

Amen

About the Author
Walter Parks

Hi! Thanks so much for your interest in my books!

My principal interests are true stories of the unusual or of the previously Unknown or unexplained. I have occasionally also written some fiction.

I was born in Memphis Tennessee and grew up in Saltillo Mississippi, a small town near Tupelo Mississippi.

After graduating from Mississippi State University as an aerospace engineer I moved to Orlando Florida and worked for Lockheed Martin for 24 years. I advanced from an aerospace engineer to a Vice President of the Company and President of the Tactical Weapons Systems Division.

I then formed Parks-Jaggers Aerospace Company and sold it 4 years later.

I continued my education throughout my career with a MBA degree from Rollins College and with Post Graduate Studies in Astrophysics at UCLA; Laser Physics at the University of

Michigan; Computer Science at the University of Florida; and Finance and Accounting at the Wharton School, University of Pennsylvania.

After selling my aerospace company I formed Quest Studios, Quest Entertainment and Rosebud Entertainment to make films at Universal Studios. I produced 10 films, directed 7 films and wrote 5 films produced at Universal Studios.

I then formed UnknownTruths Publishing Company to publish true stories of the unusual or of the previously Unknown or unexplained. These include books about past events so unbelievable that most people have relegated them to "myths".

I have published 26 books with 24 in eBook format, 15 in Paperback format and 21 as Audio Books.

Jesus the Missing Years

Atlantis the Eyewitnesses

Atlantis the Eyewitnesses Part I: Creation

Atlantis the Eyewitnesses Part II: Legacy

Atlantis the Eyewitnesses Part III: Destruction

Immortal Again

Aging is a Treatable Disease

Paranormal Portal to a Parallel Universe

Alligator Attack!

The Devil Takes the Bodies

Caribbean Ghost, Genetic Memory Comes Alive

Clan of the Bigfoot

The Body Returns, Corpus de Licti

I Look Marvelous, Skin Care Guide

Indian Massacre in Orlando

Who the Hell is Satan

Noah's Flood, the Conclusive Evidence

Jesus, His School Years

Treasure Hunt, Finding Solomon's Temple Treasure

Ancient Secrets

The Birth of Jesus, A New Christian Holiday

Finding the Soul, Surviving Death

Hormones Working for You

Cain's Wife Lilith's Daughter

Reagan's Star Wars

My Blogs:
MyUnknowntruths.wordpress.com/
Unknowntruths.wordpress.com/
Atlantiseyewitnesses.wordpress.com/
AgingIsATreatableDisease.blogspot.com/

My Websites:
UnknownTruths.com
AtlantisEyewitnesses.com
AncientSecretsBook.com
ParanormalPortalParallelUniverse.com

I have an additional 12 books in development including the following:

Aging is Preventable describes how our new knowledge of the human aging process and supplementation protocols can essentially stop aging.

End of Honor, Death of the Mafia is a true story about how the Mafia lost its honor when its members talked during the Rudy Giuliani trials.

Federal Rat describes the true story of the life and capers of a career criminal and how he became an informant and manipulated the Federal Justice System to keep getting out of prison and returning to his life of crime.

Eden Evolution addresses the questions: how did mankind really get started; was there a Garden of Eden?

Sex in the Ancient Churches describes how the ancients recognized that sex and the sun produced life and how they used both in their rituals and places of worship.

Ted Kennedy The Lion of Privilege is the true story of how Ted Kennedy and Jack Valenti misused the FBI and the IRS to close down a film production company to prevent them from making the film **Death at Chappaquiddick**.

Our Privileged Congress describes the privileges Congress has given itself without regard to benefits for the American People.

Crystal Healing describes the science of (potentially) healing crystals.

Shakma, Filming a Crazed Baboon describes the frustrating experience of making the film Shakma at Universal Studios with a crazed baboon.

How to Make a Zombie describes the science of how to make a true zombie and describes actual instances.

<u>Dam I Didn't Know That</u> describes interesting tidbits that most people do not know but are important enough to know.

<u>Alien Arrival, the First Visit</u> is a novel about alien encounters through the ages, and today.

About the Author
John Long

Many people have a calling; a purpose; a goal in their life. I never thought I had a special calling to reach a particular goal in my life until just a few years ago. That purpose, if you will, is simply to help others in some small way.

I was born in a small three room house out in the country side Northwest of Saltillo Mississippi in 1936. As a direct decent of an Irish immigrant, I grew up in an area known as "Barrett Ridge" which was named after my great Grandfather, Ned Barrett. My family was Irish Catholic and I was very devout to that religion. In Northeast Mississippi a Catholic at that time was rarer than hen's teeth.

After high school, I joined the army and served three years in the Tenth Infantry Division, spending two and a half years in Germany. Then I went to college in Memphis Tennessee and obtained two degrees from the University of Memphis, one being a law degree in 1965.

In 1968, I was elected county prosecuting attorney for Tupelo, Lee County, Mississippi and served for four years. Afterwards I practiced law in Lee County until I retired for health reasons in 1994. Most of my practice included trying cases in front of a jury. I still maintain an active license to practice for helping people pro bono where I feel needed.

My life long friend, Walter Parks, and I are writing this book about all the wonderful changes that have occurred during our lives from both our perspectives. Our universe is not the same universe we grew up with and shared in Saltillo. Some things that were true then are still true today. Some things, we have come to learn, are no longer true. The one sure thing I know is that I have a true friend in Walter Parks who wanted me to collaborate with him in writing this book. Although I have never aspired to write a book, I wanted to help in some small way.

I knew that I had made the correct decision when we met in the wilderness near Homosassa Springs in Florida and started discussing in detail how our views of life and the universe had changed during our 77 years of life.

We wanted to document what we now really believe about Life and the Universe.

I hope you enjoy our efforts and maybe it will stimulate you to consider what you really believe.

About
UnKnownTruths
Publishing Company

UnKnownTruths Publishing Company was formed to publish true stories of the unusual or of the previously Unknown or unexplained. These stories typically provide radically different views from those that have shaped the understandings of our natural world, our religions, our science, our history, and even the foundations of our civilizations.

The Company's stories also include stories of the very important anti-aging, life-extending medical breakthroughs; stem cell therapies; genetic therapies; cloning and other emerging findings that promise to change the very meaning of life.

The Company also publishes stories from the past that are so unbelievable that they are generally considered to be myths. The published stories provide the evidence for the truth.

The Company has published 26 books with 24 in eBook format, 15 in Paperback format and 21 as Audio Books.

The Company currently has an additional 12 books in development.

UnknownTruths.com